Spring Cloud
微服务快速上手

晁鹏飞 著

清华大学出版社
北京

内 容 简 介

本书介绍了当下主流的属于 Spring 生态的微服务框架——Spring Cloud，该框架继承了 Spring Boot 的优点，开发、部署等都非常简单。本书内容全面，包含了 Spring Cloud Netflix 和 Spring Cloud Alibaba 的组件，分别介绍了微服务概述、微服务注册中心、服务调用、服务的熔断降级限流、配置中心、服务网关、链路追踪、服务监控、分布式锁解决方案、分布式事务解决方案，以及微服务鉴权认证安全设计相关知识。在解读核心组件的实现原理的同时，配以案例带领读者进行实践。

本书适合需要求职进入互联网公司从事开发工作的读者、研究 Spring Cloud 技术的读者，以及从事微服务开发和对编程感兴趣的读者。

本书封面贴有清华大学出版社防伪标签，无标签者不得销售。
版权所有，侵权必究。举报：010-62782989，beiqinquan@tup.tsinghua.edu.cn。

图书在版编目（CIP）数据

Spring Cloud 微服务快速上手 / 晁鹏飞著．—北京：清华大学出版社，2022.5
ISBN 978-7-302-60482-2

Ⅰ．①S… Ⅱ．①晁… Ⅲ．①互联网络—网络服务器 Ⅳ．①TP368.5

中国版本图书馆 CIP 数据核字（2022）第 054007 号

责任编辑：贾小红
封面设计：姜　龙
版式设计：文森时代
责任校对：马军令
责任印制：丛怀宇

出版发行：清华大学出版社
　　　　　网　　址：http://www.tup.com.cn，http://www.wqbook.com
　　　　　地　　址：北京清华大学学研大厦 A 座　　　邮　编：100084
　　　　　社 总 机：010-83470000　　　　　　　　　邮　购：010-62786544
　　　　　投稿与读者服务：010-62776969，c-service@tup.tsinghua.edu.cn
　　　　　质量反馈：010-62772015，zhiliang@tup.tsinghua.edu.cn
印 装 者：三河市天利华印刷装订有限公司
经　　销：全国新华书店
开　　本：185mm×235mm　　　印　张：27.5　　　字　数：549 千字
版　　次：2022 年 6 月第 1 版　　　　　　　　　　印　次：2022 年 6 月第 1 次印刷
定　　价：108.00 元

产品编号：092540-01

前言

Spring Cloud 系列技术更新迭代的速度非常快，一直以来它的版本都是采用英国伦敦地铁站的命名方式，不过从 2020 年年底开始，它抛弃了这种命名方式，采用了日历化的版本命名方式。与此同时，Spring Cloud 移除了很多旧的组件，如 netflix-archaius、netflix-concurrency-limits、netflix-core、netflix-hystrix、netflix-ribbon 等，几乎移除了 Netflix 公司除了 Eureka 之外的所有组件，当然被移除的组件都有对应的替代方案，这些都需要我们去学习。

关于作者和本书

笔者从事软件开发工作距今已有 10 年时间，先后参与过电子政务、移动医疗、车联网、网约车等业务。从 2019 年开始，笔者从程序员转为讲师。此后，笔者接触到了大量的不同年龄、不同工作经验、不同背景的学生，从而慢慢总结出了一套让技术赋能业务的教学方式，可以让学生从解决方案的角度去学习技术。

笔者从 2015 年开始接触微服务，到现在已经有 6 年时间，其间经历了 Spring Cloud 技术的大发展，也利用 Spring Cloud 技术为公司解决了许多业务和技术问题。后来经过不断的总结，发现很多问题产生的根源是对软件原理和对框架设计的不了解导致的。所以在编写这本书时，笔者从业务场景、设计思路、落地实现、原理源码等几个方面来进行讲解，希望能给读者带来一些启发。

本书将主流的微服务解决方案基本都融合在一起。例如，注册中心既包括 Netflix 的 Eureka，也包括 Alibaba 的 Nacos，还包括 HashiCorp 的 Consul，让大家在技术选型中有一个横向的对比，可以结合自己的业务有更多的选择。另外，本书也提供了很多分布式的解决方案，如分布式锁和分布式事务，可以让大家通过对本书的学习，对微服务技术栈有一个整体的认识，同时能将所学技能应用在生产环境中。

读者对象

- ☑ 需要求职进入互联网公司从事开发工作的读者。
- ☑ 希望研究 Spring Cloud 技术的读者。

- ☑ 从事微服务开发的读者。
- ☑ 对编程感兴趣的读者。

勘误和支持

由于笔者水平有限,加之编写的时间也很仓促,书中难免会出现一些遗漏之处,恳请读者批评指正。

读者服务

读者可以通过扫描下方二维码访问本书专享资源官网、加入读者群、下载资源或反馈书中的问题。

目 录

第 1 章 微服务概述 1
- 1.1 单体架构 1
- 1.2 集群架构 3
- 1.3 微服务架构 5
- 1.4 微服务特性 6
- 1.5 微服务实践参考 9
- 1.6 微服务的缺点 11
- 1.7 Spring Cloud 简介 12
- 1.8 小结 13

第 2 章 微服务注册中心 15
- 2.1 为什么要有注册中心 15
- 2.2 注册中心的设计思路 16
  - 2.2.1 注册中心的存储结构 16
  - 2.2.2 注册中心需要具备的操作 17
- 2.3 Eureka 的使用 19
  - 2.3.1 创建注册中心服务端 Eureka Server 19
  - 2.3.2 创建客户端 22
  - 2.3.3 Eureka Server 高可用搭建 23
  - 2.3.4 Eureka Server 端用户认证 30
  - 2.3.5 自我保护机制 32
  - 2.3.6 多网卡选择 34
  - 2.3.7 Eureka Server 源码解析 34
  - 2.3.8 Eureka Client 源码解析 51
- 2.4 Nacos 的使用 57
  - 2.4.1 搭建单节点 Nacos Server 58

	2.4.2	创建 Nacos Client 60
	2.4.3	高可用 Nacos Server 搭建 61
2.5	Consul 的使用 64
	2.5.1	搭建单节点 Consul Server 64
	2.5.2	创建 Consul Client 65
	2.5.3	高可用 Consul Server 搭建 67
2.6	小结 71

第3章 服务调用 72

3.1	生产环境中的微服务架构 72
3.2	RestTemplate 调用 73
	3.2.1 RESTful 风格介绍 73
	3.2.2 RestTemplate 实战 74
	3.2.3 RestTemplate 源码解析 76
	3.2.4 负载均衡 78
	3.2.5 自定义配置负载均衡 78
	3.2.6 Ribbon 源码解析 80
3.3	OpenFeign 调用 85
	3.3.1 OpenFeign 的基础使用 85
	3.3.2 自定义 URL 87
	3.3.3 自定义 OpenFeign 的配置 87
	3.3.4 Feign 源码解析 88
3.4	小结 97

第4章 服务的熔断、降级和限流 98

4.1	熔断和降级的应用场景 98
4.2	熔断和降级的使用 100
	4.2.1 RestTemplate 中熔断和降级的使用 101
	4.2.2 OpenFeign 中熔断和降级的使用 102
4.3	自定义熔断配置 102
4.4	限流 104
	4.4.1 计数器（固定窗口）算法 104
	4.4.2 滑动时间窗口算法 105

4.4.3　漏桶限流算法 106
　　4.4.4　令牌桶限流算法 106
4.5　Sentinel 熔断和限流实战 107
　　4.5.1　Sentinel 控制台安装 107
　　4.5.2　Sentinel 在程序中的配置 109
　　4.5.3　Sentinel 流控规则 113
　　4.5.4　Sentinel 降级规则 116
　　4.5.5　Sentinel 热点规则 117
　　4.5.6　自定义流控处理 118
4.6　小结 119

第 5 章　配置中心 120
5.1　配置中心应用场景 120
5.2　配置中心的设计思路 120
　　5.2.1　配置存储 120
　　5.2.2　配置的属性 121
　　5.2.3　配置服务 122
5.3　Spring Cloud 配置中心的使用 123
　　5.3.1　在 Git 上创建配置 123
　　5.3.2　创建配置的服务端 123
　　5.3.3　创建配置的客户端 126
　　5.3.4　配置的手动刷新 130
　　5.3.5　配置的自动刷新 131
　　5.3.6　在 MySQL 上创建配置 134
　　5.3.7　配置内容对称加密 135
　　5.3.8　配置内容非对称加密 138
　　5.3.9　配置中心安全认证 141
　　5.3.10　高可用配置中心 142
5.4　Nacos 配置中心使用 142
　　5.4.1　Nacos 配置中心的基本使用 143
　　5.4.2　Nacos 配置扩展 145
　　5.4.3　Nacos 模型管理 147
5.5　小结 147

第 6 章 服务网关 148
6.1 网关 Gateway 的基本使用 148
6.1.1 微服务搭建 passenger-api 148
6.1.2 Gateway 网关搭建 cloud-gateway 150
6.1.3 Java 类加载器层级结构 152
6.1.4 Java 双亲委派机制原理 154
6.1.5 Java ClassLoader 类的原理 155
6.1.6 Java URLClassLoader 类的原理 156
6.1.7 Java 双亲委派机制的打破 159
6.1.8 Java 自定义类加载器 160
6.2 路由断言使用 171
6.2.1 Path 路由断言 171
6.2.2 Query 路由断言 172
6.2.3 Method 路由断言 173
6.2.4 Header 路由断言 174
6.2.5 自定义路由断言 175
6.3 过滤器的使用 181
6.3.1 添加请求头过滤器 181
6.3.2 移除请求头过滤器 182
6.3.3 状态码设置 185
6.3.4 重定向设置 185
6.3.5 过滤器源码 186
6.4 全局过滤器 187
6.5 小结 189

第 7 章 链路追踪 190
7.1 链路追踪的设计思路 190
7.2 链路追踪的使用 191
7.3 追踪原理分析 193
7.4 可视化链路追踪 194
7.5 消息队列收集链路追踪 199
7.6 小结 200

目　录

第8章	服务监控	201
8.1	Spring Boot Admin 的使用	201
8.2	监控内容介绍	206
8.3	认证保护	207
8.4	服务监听邮件通知	208
8.5	服务监听钉钉通知	210
8.6	小结	220
第9章	分布式锁解决方案	221
9.1	业务场景	221
9.2	单机 JVM 锁	224
	9.2.1　系统架构与核心代码	224
	9.2.2　JMeter 安装与配置	228
	9.2.3　压力测试	232
	9.2.4　单机 JVM 锁的问题	235
9.3	分布式锁思路分析	238
9.4	MySQL 分布式锁	239
9.5	Redis 分布式锁	242
	9.5.1　死锁问题	242
	9.5.2　过期时间问题	244
	9.5.3　Redisson 框架使用	246
	9.5.4　Redis 单节点问题	248
	9.5.5　红锁	248
	9.5.6　Redis 做分布式锁的终极问题	255
9.6	Zookeeper 分布式锁	255
	9.6.1　Zookeeper 节点类型	256
	9.6.2　Zookeeper 分布式锁原理	256
	9.6.3　Zookeeper 结合 MySQL 乐观锁	257
	9.6.4　Zookeeper 分布式锁代码实现	258
9.7	小结	260
第10章	分布式事务解决方案	261
10.1	分布式事务业务场景	261

VII

10.2	分布式事务思路分析	264
10.3	X/Open 分布式事务模型	265
10.4	两阶段提交协议	267
	10.4.1 两阶段提交协议的过程	267
	10.4.2 两阶段提交协议的缺点	268
10.5	三阶段提交协议	269
	10.5.1 三阶段提交协议的过程	269
	10.5.2 两阶段提交协议和三阶段提交协议的区别	271
10.6	CAP 定理和 BASE 理论	271
	10.6.1 CAP 定理	272
	10.6.2 BASE 理论	273
10.7	TCC 分布式事务解决方案	274
	10.7.1 TCC 方案	274
	10.7.2 TCC 方案的异常处理	276
10.8	可靠消息最终一致性方案	278
	10.8.1 可靠消息最终一致性问题分析	278
	10.8.2 本地消息事件表方案	279
	10.8.3 RocketMQ 事务消息方案	281
10.9	RocketMQ 安装部署	282
10.10	RocketMQ 事务消息实战	286
	10.10.1 生产者 producer	286
	10.10.2 消费者 consumer	309
10.11	Seata 分布式事务解决方案	326
10.12	Seata AT 模式实战	328
	10.12.1 启动注册中心	329
	10.12.2 下载安装 Seata	329
	10.12.3 搭建订单服务	335
	10.12.4 搭建库存服务	351
	10.12.5 测试	365
10.13	Seata TCC 模式实战	368
	10.13.1 订单服务	368
	10.13.2 库存服务	372

		10.13.3 测试	375
10.14	最大努力通知方案		379
	10.14.1	什么是最大努力通知方案	379
	10.14.2	最大努力通知方案实战	380
10.15	小结		414

第11章 微服务鉴权认证安全设计 ... 415

11.1	鉴权认证常见的场景及解决方案		415
	11.1.1	单体应用	416
	11.1.2	微服务应用	416
11.2	OAuth 2.0 介绍		417
11.3	OAuth 2.0 实战		419
11.4	JWT 使用		425
	11.4.1	JWT 的介绍	425
	11.4.2	JWT 的实践	426
	11.4.3	JWT 的使用场景	428
11.5	小结		428

第 1 章 微服务概述

互联网始于1969年美国的阿帕网（ARPA），最开始的阿帕网只在美国军方使用。随着时间的推移，一些大学也开始加入建设，慢慢演化成了现在的因特网（Internet）。随着计算机网络的普及，到现在全世界几乎一半的人口，都在使用互联网产品。日常生活中的各种场景，如商场购物、沟通交流、金融理财、货运物流等，都可以在网上实现。随着网上的应用越来越多，用户也越来越多，业务场景也越来越复杂，传统的单体应用已经无法满足互联网技术的发展要求。随着业务复杂度的逐渐提升，代码的可维护性、可扩展性和可阅读性在逐步降低，修改和更新代码的风险也变得越来越高。而技术作为业务的支撑，它永远伴随着业务的发展而发展，所以为了解决单体应用的缺点，诞生了微服务。

最近几年，微服务很受欢迎，无论是在公司的实际开发应用中，还是在技术人员之间的日常话题交流中，微服务的技术都是主流。微服务概念是著名的IT专家Martin Fowler提出来的，它是用来描述将软件应用程序设计为独立部署服务的一套思想，一般按照业务进行服务划分，它有自动化运维、容错、快速迭代的特点，给软件领域带来了巨大的影响。

那么，微服务是怎么一步一步演进到现在的？在实际生产开发中，如何正确使用微服务呢？这些问题一定要先搞清楚，否则做微服务项目，就会做成大型的分布式单体应用。要知道任何技术和方案的形成都不是一蹴而就的，很多思想、解决方案都是在原有理论的基础上进一步发展的。因此我们先来回顾一下软件架构的发展历程，了解这些知识会对以后的架构选型有很大的帮助。

1.1 单体架构

在传统的软件架构中，经常提到MVC（model-view-controller）模型，也就是模型视图控制器。即M（model）数据访问层、V（view）视图层、C（controller）业务逻辑控制层。

☑ 模型层：即数据访问层，是业务逻辑控制器层的支撑部分，它可以向业务逻辑控制器层提供数据来源，也可以将业务逻辑控制器层处理的结果数据进行持久化。

☑ 视图层：用于软件和用户的交互。例如，接收用户的输入信息，将响应信息呈现给用户。一般应用都是网页、App、小程序。

☑ 业务逻辑控制器层：即业务逻辑处理层，将用户输入的信息，根据一定的业务逻辑进行加工处理。

开发服务端的软件一般都是按照 MVC 架构来设计的，其主要功能包括响应用户的交互、对业务流程的操作、对数据库的增删改查。虽然有三层模型的划分，但这只是在软件架构上的分层，并没有对业务场景进行划分。一个单体系统就是将所有业务场景的视图层、业务逻辑控制器层、模型层全部囊括在一个工程内，所有的功能都在一个项目中开发，开发完成后打成 jar 包，或者 war 包，最终部署到一台服务器上。单体应用架构如图 1-1 所示。

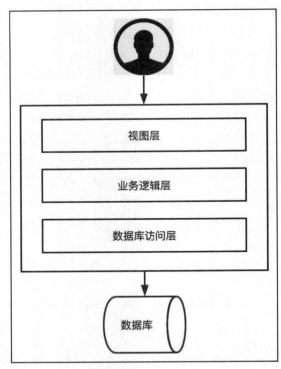

图 1-1 单体应用架构图

在互联网早期，由于业务比较简单，访问量比较少，开发人员采用单体应用会有很多优点，下面从开发、测试、部署运维的角度分别进行分析。
- ☑ 易于开发。一般在 IDE（integrated development environment，集成开发环境）中新建一个项目就可以进行开发，开发完成后进行测试，测试通过后可以直接打包，即可部署。
- ☑ 易于测试。不需要其他的服务配合，直接独立运行起来，就可以进行测试。
- ☑ 易于部署。只有一个 jar 包，或者一个 war 包，也可以将程序中用到的文件资源、数据库等直接部署到服务器上，只要服务器拥有相关的运行环境即可。

1.2 集群架构

随着业务的发展，系统的压力越来越大，对设备的配置要求也越来越高，工程师们不断上线性能更高的服务器，而高端点的计算设备又非常昂贵，如 IBM 的 Power 系列小型机，一台就需要十几万美元，这对于小型企业来说是不能承受的，如阿里巴巴的淘宝就经历过"去 IOE"的过程，"去 IOE"是阿里巴巴公司提出的概念，其本意是在阿里巴巴的 IT 架构中，去掉 IBM 的小型机、Oracle 的数据库、EMC 的存储设备，代之以自己公司在开源软件基础上开发的系统。这其中的原因包括设备购买的费用昂贵、设备的维护需要专业的人员，以及维护的服务费也比较高。并且，由于有摩尔定律，处理器的性能每隔两年翻一倍，单台机器性能发展得再快，也跑不过业务的指数级发展。

单体应用的部署可能还会造成硬件资源的浪费。例如，在用小型机部署系统时，系统在 1 月份只需要 10GB 内存，预计在"双 11"时会达到 100GB 内存的需求，这样企业就需要为将来的业务预留 100GB 的内存配置，如此就会造成在一段时间内有 90GB 内存的浪费。如果到"双 11"时，业务真正达到了 100GB 内存的需求还好，没有浪费内存，如果达不到呢，90GB 内存就白白浪费了。

基于以上提到的两点问题，能想到的解决方案如下。
- ☑ 机器昂贵。通过分解化整为零，使高配的机器由众多低配的机器来组成，从而解决成本问题。
- ☑ 资源浪费。将配置进行适时的增加。在合适的时候，再采用合适的配置。如到"双 11"的时候，再进行配置的增加。

当时市场上有很多价格便宜的普通 PC，和小型机相比而言，虽然普通 PC 的 8GB 内存比不上小型机的 128GB 内存的性能，但是两者的计算架构是一样的，能实现的计算功能也是一样的，而 PC 的价格却要比小型机便宜很多。于是聪明的工程师们又想出了一种廉价的解决方案，将很多普通 PC 集群到一起，这就是集群架构，这样一来，一方面可以将小的力量汇聚成大的力量，大大降低了采购小型机的成本；另一方面当需要增加配置时，再增加机器即可，即减少了资源闲置和浪费。后来大多数公司都采取了集群架构进行系统部署，通过增加负载均衡服务器，如硬件负载均衡 F5、软件负载均衡 Nginx 等来进行系统架构的搭建。

负载均衡往往对外暴露一个统一的接口，让外界用户感觉是在向一台服务器发起请求，而实际上通过负载均衡已经将请求分发到了后端不同的服务器上。并且负载均衡还可以进行服务器的健康检查，当有新机器加入时，可以将流量进行动态地分配；当有服务器宕机时，也可以从负载均衡中将其剔除。这样一来，如果想让服务负载原来 2 倍的访问量，只需要横向、水平扩容 1 倍的服务器即可。

由于有了统一的入口，因此可以在这个入口进行流量清洗，可防 DDoS 攻击。

这种集群的思想不仅适用于服务器，也适用于缓存和数据库。例如，有的系统为了提高查询效率，还增加了缓存服务器（主从备份）；有的系统为了缓解数据库压力，进行数据库的读写分离（一主多从）。通过这些方法可应对用户量的增加对系统带来的压力，此时的集群架构部署如图 1-2 所示。

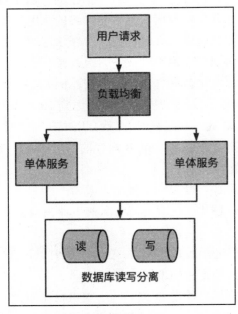

图 1-2 集群架构部署

集群架构虽然具备一定的处理高并发的能力，改善了系统的性能，但是从业务的角度来看，它依然是部署在不同机器上的一模一样的服务，其本质上还是一个单体的架构，所有的功能都糅合在一个项目中。随着业务的发展，软件服务的功能越来越复杂，物极必反，它的优点随着软件行业的发展，也变成了它的缺点，集群架构的弊端逐渐明显。

- ☑ 开发风险高。随着业务越来越复杂，代码量越来越大，在以后需要升级系统时，即使改动一行代码，也需要重新编译部署整个项目，如果改动还影响到了其他模块，则会有比较大的隐患，因此改动代码的风险非常高。
- ☑ 测试成本高。由于都在一个服务中，不确定改动是否会对其他功能造成影响，所以每一个小小的改动，都需要做全量测试，从而额外增加了测试的成本。
- ☑ 部署频率低。由于开发的风险高，需要测试的成本高，从而导致更新频率变慢，所以不能经常部署，而现在软件更新迭代的频率又非常高，这样就无法满足对客户快速交付的需求。

当然还有很多问题。例如，由于所有功能都在一个项目中，不同功能对硬件的要求也是不同的，有的功能处理文件较多，是 IO 密集型，有的功能计算功能较多，是 CPU 密集型，如果在单体应用中，它们又都是在一个项目中，就无法针对不同的功能，进行硬件资源的合理分配，这样就会造成一定程度的硬件资源的浪费。

另外，如果通过负载均衡部署了 10 个服务，而每个服务里面都有用户登录功能，当需要对用户登录功能进行升级时，这 10 个服务也都需要同时升级，这种重复建设显然是一种浪费。

在软件开发领域，重复建设是不可取的，为了解决这些问题，微服务架构应运而生。

1.3 微服务架构

为了解决单体应用的缺点，工程师们想到将原来大的单体应用进行拆分，化整为零形成独立的应用，不过此时这些应用没有直观的入口，因此用传统应用的概念来定义就不太妥当。于是诞生了"服务"，通过服务来描述这种功能性的应用，并为其他应用提供功能支持，服务于其他应用。现在"服务"这个词已经广义化了，只要是能向其他组件提供技术支撑的系统都叫服务，甚至出现了很多概念，如 SaaS（软件即服务）、PaaS（平台即服务）、IaaS（基础架构即服务）。

服务其实就是将其他应用当成消费者，为其提供特定的功能。例如，将用户登录抽象成一个服务，可以叫作用户服务，让其他服务来调用。以此类推，可以有订单服务、支付服务、消息服务、商品服务、客服服务、营销服务、积分服务等。总之，就是抽取公共部分以避免重复工作。这样一来，后续应用的开发，也就变成了服务的开发。

理解了服务是怎么回事，接着我们来学习微服务。

微服务概念最早是 Martin Fowler（马丁·福勒）于 2014 年提出的。他指出了微服务架构是用于描述将软件应用设计成独立部署服务的一种方式，但是并没有给出精确的定义，只是提出了一些确定、通用的特征，如围绕组织和业务能力、自动化部署、节点智能、对语言和数据的分散控制等。

从技术的角度理解，微服务就是将传统的一站式应用，根据不同的业务拆分成一个小的服务，每个服务提供独立的业务功能，拥有自己独立的存储（数据库或者缓存），通过服务之间的互相调用来完成复杂的系统功能。

1.4 微服务特性

既然系统采用了微服务架构，就需要了解一些微服务的特性，这样在进行微服务开发时，脑海中才会有一些指导方向。微服务具有以下特性。

1. 服务组件化

组件是独立、可替换、可升级的软件的单元。将整体应用拆分成独立的服务组件后，当对单个组件的修改完成后，只需要重新部署该组件即可。这也不是绝对的，有一些改动会影响服务间的接口，调用方（消费者）也需要进行相应的修改与部署。所以好的微服务架构的目标是通过高内聚、低耦合来尽量减少不同服务之间的依赖。

2. 围绕业务能力组织团队

康威定律指出：设计系统的架构，受制于产生这些设计的组织的沟通结构。不同公司的产品结构，基本上反映了组织的沟通结构，如图1-3所示。

在一个单体应用中，一个项目会有很多研发团队的人参与，如前端工程师组、后端工程师组、数据库管理员组，不同组的人参与同一个项目，彼此沟通成本很高。单体项目的分工图如图1-4所示。

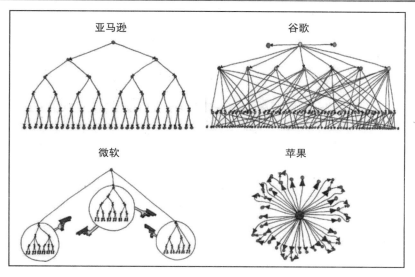

图 1-3　不同公司的组织沟通结构图

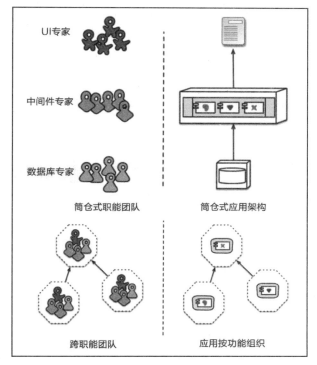

图 1-4　单体项目分工图

在微服务项目中，团队是围绕业务来进行组织的。例如，某个服务组里有前端工程师、后端工程师、数据库管理员，这算是一个工作单元，能极大降低沟通成本。

3．产品不是项目

在大部分软件开发时，采用的都是项目的形式，开发团队开发完软件后将软件移交给运维团队去维护，原来的项目团队解散。微服务支持者倾向于避免这种模式，他们认为团队应该在产品的整个生命周期内对它负责。这就是 Amazon（亚马逊）提出的"you build, you run it"。这么做的好处是可以让开发人员经常接触到他们的产品，从而了解产品在生产中的行为，并增加与用户的联系，同时在业务升级时也很方便，毕竟开发和升级都是由同一个团队负责。

4．去中心化的数据管理

在单体应用中，数据都在一个数据库（这里指的是逻辑库，可能物理上部署多个数据库服务器）中，管理难度大。在微服务中，不同的业务数据是放到不同的服务所对应的数据库中的。如订单存放到订单库、商品存放到商品库等，从而大大降低了数据的管理难度。去中心化数据库示例如图 1-5 所示。

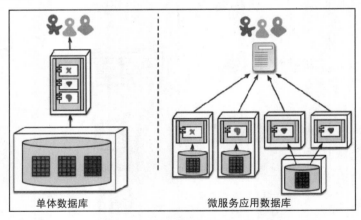

图 1-5　去中心化数据库示例

5．分散治理

由于将大服务拆分成了小服务，所以每个服务都可以根据自己的需求，采用合适的技术实现方案，如采用不同的语言（有的业务适合用 Go，有的业务适合用 Java）、采用不同的存储（有的业务适合用 Redis，有的业务适合用 MySQL）等。这样更有利于提高开发效率，节省资源。

了解了微服务的特性，作为一名有着架构师梦想的程序员，应该清楚地认识到微服务可以给实际工作带来的好处。

- ☑ 从公司团队管理角度。将服务拆分成微服务后，可以让独立的团队完成独立的服务，实现不同的业务，使得团队更聚焦，减少了跨部门沟通的障碍，大大提高了开发效率。
- ☑ 从公司产品研发角度。由于服务彼此独立，可以针对不同的产品需求，针对不同的服务快速地进行产品升级。从而在快节奏的软件开发环境中，缩短对市场的响应时间，加快产品的迭代。

1.5 微服务实践参考

了解了微服务的好处，那么如何在实际生产中进行微服务的拆分？微服务中的微到底如何界定？服务拆分后如何进行协作？下面从开发工程师的角度来和大家聊一聊在实际开发中如何进行微服务的实践。

1. 服务如何拆分

微服务的拆分标准没有一个特别明确的概念，记住一条原则即可，就是不管服务如何拆分，都要明确定义好微服务系统的边界。举个例子，在电商系统中，可以有订单服务、用户服务、支付服务。订单服务还可细分为订单列表服务、订单详情服务等。这些微服务的系统边界需要开发团队自己来决定。

这样做的好处是，将一个大的业务拆分成了若干个小的业务，每个业务就是一个服务，将复杂业务简单化，编码也会相应的简单，代码的可读性和可扩展性都会增强。如果团队加入新人的话，也能减少学习成本。

2. 服务间的通信

服务拆分好之后，完成了开发就可以部署到独立的服务器上了。而服务之间是需要通信的，一般倾向于使用 HTTP 协议通信，大部分采用 RESTful 风格，这种通信机制与平台和语言无关，因此可以调用跨语言的微服务。例如，可以调用 Go 语言编写的微服务，也可以调用 Java 语言编写的微服务。

服务之间还可以通过消息中间件进行通信。例如，服务 A 将消息发送到消息中间件，服务 B 通过订阅消息中间件进行后续的业务处理。

由于这种无状态的通信方式使得服务与服务之间没有任何的耦合，所以随着业务的发展，可以将服务更进一步细分，只要增加服务之间的调用接口即可。如果并发量继续增加，还可以像前面一样，将微服务做成集群，从而提高系统的负载能力。

3．每个服务的数据库分别独立

在单体应用中，由于只有一个服务，所以业务共用一个数据库就行。而随着业务量的增加，数据库的表越来越多，就越来越难以管理和维护，数据库的性能也越来越慢。如果按照业务来拆分，这样数据库也就对应的独立了。每个微服务都有自己对应的数据库，这相当于将原来一个数据库的压力分散到多个数据库上。每个数据库的关系也会变得简单，开发也简单，数据库的性能也会有所提高。

大厂面试

面试官：你在公司中做微服务项目时，是依据什么来进行服务划分的？微服务设计一般都遵循了什么原则？

面试者：在进行微服务设计时，服务的数量相对于单体应用来说会比较多。考虑的重点就是如何准确识别系统的隔离点，也就是系统的边界。只有每个服务的边界确定了，才能在以后的开发中做到更好地协作。识别系统的隔离点，需要遵守下面几个原则。

（1）单一职责原则。让每个服务能独立、有界限地工作，每个服务只关注自己的业务，做到高内聚，服务和服务之间做到低耦合。

（2）服务自治原则。每个服务要能做到独立开发、独立测试、独立构建、独立部署、独立运行，与其他服务进行解耦。

（3）轻量级通信原则。让每个服务之间的调用是轻量级的，并且能够跨平台、跨语言调用。如采用 RESTful 风格、利用消息队列进行通信等。

（4）粒度进化原则。对每个服务的粒度把控，其实没有统一的标准，这个需要结合具体业务问题去确定。不要过度设计，服务的粒度随着业务和用户的发展而发展。

总结一句话，软件是为业务服务的，好的系统不是设计出来的，而是进化出来的。

1.6 微服务的缺点

事务总具有两面性，微服务在有很多优点的同时，也有很多缺点。

1. 增加了系统的复杂度

将单体应用拆分成微服务后，开发人员要花时间和精力去学习更多的框架知识。网络调用频繁会带来延时、出错的风险。虽然说服务之间是解耦的，但是如果修改一个服务影响到了其他服务的调用，就会带来不必要的麻烦。

2. 分布式一致性

在微服务开发中，由于是很多个微服务来协作完成某一个业务，需要系统间的调用来实现业务逻辑，在一个服务里只有一个数据库时，可以通过数据库的事务来保证数据的一致性。有一个著名的 CAP 定理，即一致性、可用性和分区容错性，这三个条件同时满足是不可能的，最多能满足其中的两个。由于分布式微服务系统需要服务之间通过网络进行调用，所以必须满足 P（这个内容在后面的章节中介绍），即分区容错性。现在是多个服务多个库，所以数据的一致性和可用性没办法同时满足，要么是 CP，要么是 AP。一般微服务为了可用性，往往会牺牲一致性。

3. 增加系统开销

服务调用会涉及服务间地址的管理，需要额外的服务治理组件；服务调用失败会涉及熔断降级的处理，需要熔断降级的组件。由于服务多，更多的服务意味着更多的运维投入，部署起来复杂度也会提高，需要自动化运维，服务健康状态的监控也会变得复杂。当服务调用出现问题时，错误的排查也比单体应用要复杂，会涉及链路追踪等技术。

总之，单纯服务数量的增加，不仅是服务的增加，而是对整个业务体系提出了更高的要求。给大家讲一个小例子，便于大家理解。

有一家小饭店，起初只有一个厨师，切菜、洗菜、备料、炒菜这些工作全都是他一个人做。后来客人渐渐变多了，一个厨师肯定忙不过来，所以老板又请了一个厨师，这两个厨师都能炒一样的菜，可见这两个厨师的关系是集群。

突然有一天，一下来了1万个人吃饭，让1万个人同时进门，需要饭店多开几个门；为了能同时做好1万个人的饭菜，需要扩大厨师团队，将原来的做菜步骤进行分工协作，需要切菜服务、洗菜服务、备料服务、炒菜服务，如果某个服务出问题了，要能快速定位问题并进行补救；为了能同时将菜送到1万个人的桌子上，需要进行路径计算，不能浪费上菜时间；甚至排水系统都得重构等。

所以，对比来看，吃饭人数的增加不是单纯靠多雇几个厨师就可以解决的，它需要的是一整套解决方案。

由于服务的增多，服务之间的调用必然增多。由于存在分布式网络的八大谬误，服务之间的网络调用会极大地增加系统的复杂度。为了解决复杂的网络调用，人们提出了服务网格（service mesh）的概念，服务网格将网络出错的解决方案从业务代码中剥离，沉淀到了边车（side car）中，从这个角度来说，可以把服务网格理解成下一代微服务，这是微服务未来的发展方向。

分布式网络八大谬误如下。

（1）The network is reliable（网络是可靠的）。
（2）Latency is zero（反应时间为零）。
（3）Bandwidth is infinite（带宽是无限的）。
（4）The network is secure（网络是安全的）。
（5）Topology doesn't change（拓扑不会改变）。
（6）There is one administrator（这里会有一个系统管理员）。
（7）Transport cost is zero（传输代价为零）。
（8）The network is homogeneous（网络是均匀的）。

1.7　Spring Cloud 简介

采用微服务会带来更清晰的业务划分和更好的可扩展性，在很多企业中十分流行。支持微服务的技术栈也是多种多样。当前主流的是 Spring Cloud 和 Dubbo，简单做一下对比，如表1-1所示。

表 1-1　Spring Cloud 和 Dubbo 对比

	Spring Cloud	Dubbo
注册中心	Eureka，Nacos，Consul，ETCD，Zookeeper	Zookeeper
服务间调用方式	RESTful API 基于 HTTP 协议	RPC 基于 Dubbo 协议
服务监控	Spring Boot Admin	Dubbo-Monitor
熔断器	Spring Cloud Circuit Breaker	不完善
网关	Zuul，Gateway	无
配置中心	Config，Nacos	无
服务追踪	Sleuth+Zipkin	无
数据流	Spring Cloud Stream	无
批量任务	Spring Cloud Task	无
消息总线	Spring Cloud Bus	无

从表 1-1 的比较中可以看出，Spring Cloud 的功能比 Dubbo 更全面、更完善，并且作为 Spring 的旗舰项目，它可以与 Spring 的其他项目无缝结合，完美对接，整个软件生态环境比较好。Spring Cloud 就像品牌机，整合在 Spring 的大家庭中，并做了大量的兼容性测试，保证了机器各部件的稳定。Dubbo 就像组装机，每个组件的选择自由度很高，但是如果你不是高手，如果你选择的某个组件出了问题，就会导致整个机器的宕机，造成整体服务的不可用。

> **大厂面试**
>
> 　　面试官：请回答一下，你在工作中是如何进行微服务技术选型的？Spring Cloud 和 Dubbo 有什么区别？
>
> 　　分析：结合上面的表 1-1 中的比较，和下面品牌机和组装机的例子来回答。

在本书中主要讲解 Spring Cloud 技术栈，会通过业务场景驱动的方式来学习 Spring Cloud 的各个组件，包括流行的 Spring Cloud Netflix 和 Spring Cloud Alibaba，以及其他流行的组件在微服务中的使用。

1.8　小　　结

本章作为微服务的概述，从软件架构发展的历程开始，首先向大家介绍了单体应用，分析其优点和缺点；为了解决单体应用的性能问题，采用了集群架构，然后随着业务越来越复杂，原有架构暴露出了各种开发风险高、测试成本高、部署频率低的问题，然后引入了微服务架构，也对微服务带来的优缺点进行了阐述，便于在实际工作中进行架构的取舍。另外，也对微服务发展的未来进行了展望。

　　虽然微服务没有明确的定义，但是马丁·福勒提出了微服务的一些特性，在设计微服务架构时，应尽量考虑到这些特性，将这些特性作为指导实践的思想。最后，站在开发的角度对微服务实践的方案进行了阐述，并且在微服务的实现方案中，对主流的微服务框架 Spring Cloud 和 Dubbo 进行了对比，可用来作为大家在做技术选型时的一个依据。

第 2 章 微服务注册中心

在项目开发中采用微服务设计时,由于将原来的单体应用拆分成了微服务,当要实现一项业务功能时,需要调用多个服务。这时会引发另外一个问题,多服务之间互相调用,需要知道对方的 URL(uniform resource locator,统一资源定位器),至少需要知道被调用方的 IP 地址和端口。当服务很多时,如一些大型的电商系统有成千上万个服务,每个服务有各自的 URL,这么多的 URL 不可能用硬编码的方式写到代码中,这样不利于维护。那应该如何管理呢?这时候就需要注册中心功能。

2.1 为什么要有注册中心

试想一下这个场景,如果要对某个服务进行迁移,如从北京机房迁移到杭州机房,导致它的 IP 有了变动,这就需要去更改代码(或者配置文件)中的 URL。如果变动的服务数量很多,在每个服务里都要对它的配置进行修改,修改代码(或者配置文件)的工作量就会很大,如果手一抖,敲错了 IP 中的数字,服务调用就会出问题。

该如何解决这个问题呢?其实软件行业中的很多解决方案都来源于实际生活。大家想一下,一个服务调用多个服务,是不是类似于一个人给多个人打电话。如果要记住多个人的电话号码,是不是很难?并且容易记错。那怎么办呢?答案是使用通讯录。

例如,张三要给李四打电话,有以下两种情况。

(1)没有通讯录。张三需要在自己的脑海中记住李四的电话号码,当需要给李四打电话时,张三回忆起号码(记忆力强,没记错号码)后,拿出手机拨号。

(2)有通讯录。张三无须记住李四的电话号码(张三的大脑好轻松),当需要给李四打电话时,张三先去通讯录查一下李四的号码,查到号码后拿出手机拨号。当李四的电话号码有更改时,只需要把通讯录中李四的号码进行更正即可。

其实微服务之间的调用，和打电话时利用通讯录是一样的。下面来具体学习注册中心是怎么设计的。

2.2 注册中心的设计思路

通过微服务和通讯录的类比，如果要设计一个注册中心，应该如何设计呢？程序是数据结构和算法的结合，那么要设计一个注册中心，要从下面两个方面来考虑。
- ☑ 存储结构。注册中心要存储哪些数据？怎么存？
- ☑ 提供的操作。注册中心可以提供哪些操作供服务使用？

2.2.1 注册中心的存储结构

还是类比一下通讯录。生活中的通讯录，存储的信息如表 2-1 所示。

表 2-1 通讯录示例

姓　名	电　话
张三	139××××××××

同理，注册中心是不是也应该这样存储信息呢（见表 2-2）？

表 2-2 注册中心示例

服　务　名	服　务　信　息
order-service（订单服务）	服务的 IP 地址、服务端口、服务对外提供的 URL 等

那么在程序中如何存储信息呢？通过上面的存储内容和存储方式是不是能想到 Java 中常用的数据结构 Map？我们打开 Eureka 的源码印证一下。在 AbstractInstanceRegistry.java 中有一个属性，代码如下。

```
private final ConcurrentHashMap<String, Map<String, Lease<InstanceInfo>>> registry
    = new ConcurrentHashMap<String, Map<String, Lease<InstanceInfo>>>();
```

该属性的存储结构用的是 ConcurrentHashMap，它的键值是服务名，值是一个 Map。而值的 Map 中键是服务实例 ID，值是租约（租约里面包括服务信息）。注册表的数据存储结构如图 2-1 所示。

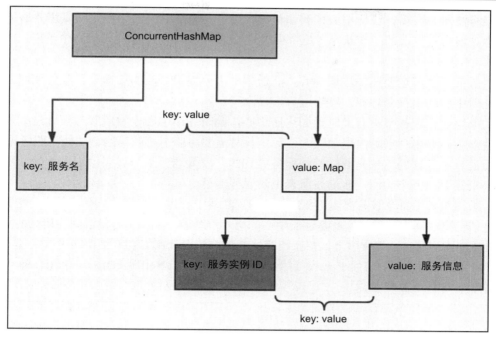

图 2-1　注册表数据存储结构

大厂面试

面试官：为什么套用两层 Map 呢？

面试者：因为在微服务系统中，为了避免服务的单点故障，通常使用的都是服务集群，一个服务名会对应多个服务实例。所以在每个服务实例的存储结构外面，又包了一层 Map，这样一个服务名就可以对应多个服务实例了。

2.2.2　注册中心需要具备的操作

通过上面的分析，我们知道了服务的注册信息如何存储。那么需要对服务的注册信息进行哪些操作呢？

我们想一下，服务启动后是需要向注册中心进行注册的，所以注册中心需要提供一个接口，让服务调用该接口来进行服务的登记注册，这就是注册中心的第一个功能，接收服务注册。

服务注册完成后，注册中心需要知道这个服务是否还是有效服务？所以需要服务定期地告诉注册中心自己的工作状态（是否可用）。此时需要注册中心提供第二个功能，接收服务心跳。

当服务下线时也要通知注册中心，注册中心需要提供对应的接口来让服务调用。此时需要注册中心的第三个功能，接收服务下线。

如果服务出现故障没有及时通知注册中心，此时注册中心也发现服务最近没有发送心跳，注册中心要主动剔除出现故障的服务。此时需要注册中心的第四个功能，服务剔除。

注册表中存储的信息是要供其他服务查询的，就像通讯录一样要供主人查阅，所以注册中心还需要第五个功能，查询注册表中的服务信息。

一般微服务中每个服务都要避免单点故障，注册中心也要做集群，所以还涉及注册中心之间注册信息的同步问题。这就是注册中心的第六个功能，注册中心集群间注册表的同步。

在 Eureka、Nacos 等官网上，都有相应 API 的介绍，调用 HTTP 接口就可以实现上面的操作。以 Eureka 为例，访问 Eureka 官网 https://github.com/Netflix/eureka/wiki/Eureka-REST-operations，如图 2-2 所示。

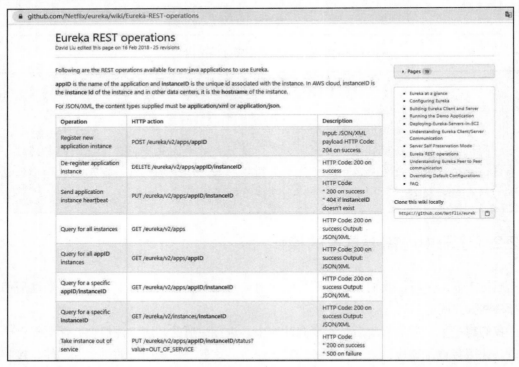

图 2-2　Eureka 官网

通过以上介绍，基本上将注册中心的需求分析清楚了。其实注册中心的本质就是一个 Web 服务，通过提供上面分析的 6 个接口供服务调用。这就是注册中心服务端，而对应的调用注册中心的服务（业务服务）一般称为注册中心客户端。

使用注册中心分为如下两步操作。

（1）搭建注册中心服务端。

（2）启动业务服务（即注册中心客户端），让服务和注册中心连通。

下面演示主流的注册中心的使用方法。

大厂面试

面试官：你能讲一下注册中心的原理吗？

面试者：在回答这个问题时，先要了解注册中心要解决的问题是什么？它主要是进行服务信息的管理，如注册表（包括服务实例的信息）的管理、服务上线、服务有效性的维持（心跳）、服务下线、无效服务的剔除、注册中心集群间的注册表同步，以及能够向注册中心客户端提供注册表信息的查询。

知道了它要解决的问题之后，再来对它进行设计，设计包括数据的存储，以及对数据的操作两个方面。存储是通过 Map 数据结构来完成的（此时将上面讲的注册中心存储结构向面试官画一下）。然后是数据的操作（将上面讲的注册中心需要具备的操作向面试官描述一下）。

2.3　Eureka 的使用

Eureka 是 Netflix 开发的一款服务发现框架（其实就是注册中心），它作为组件被 Spring Cloud 集成到了其子项目 spring-cloud-netflix 中，用来在 Spring Cloud 家族中提供服务注册发现功能。

这里使用的软件版本分别是 JDK 1.8、Spring Boot 2.4.3、Spring Cloud 2020.0.1。接下来按照下面的步骤进行 Eureka Server 的搭建。

2.3.1　创建注册中心服务端 Eureka Server

先创建一个 Spring Boot 项目 eureka-server-single。其实大部分的 Spring Cloud 组件都是

按照如下 3 个步骤来操作的。

（1）在 pom 文件中添加如下依赖。

```xml
<dependencies>
    <dependency>
        <groupId>org.springframework.boot</groupId>
        <artifactId>spring-boot-starter-web</artifactId>
    </dependency>
    <!--eureka server jar包-->
    <dependency>
        <groupId>org.springframework.cloud</groupId>
        <artifactId>spring-cloud-starter-netflix-eureka-server</artifactId>
    </dependency>

    <dependency>
        <groupId>org.springframework.boot</groupId>
        <artifactId>spring-boot-starter-test</artifactId>
        <scope>test</scope>
    </dependency>
</dependencies>
```

（2）配置 application.yml 文件。

```yaml
server:
  port: 8761

eureka:
  instance:
    hostname: localhost
  client:
    # 是否向 Eureka Server 注册，因为此时本服务就是一个 Eureka Server，其实就是配置是否向自己注册
    registerWithEureka: false
    # 是否从 Eureka Server 拉取注册表
    fetchRegistry: false
    serviceUrl:
      defaultZone: http://${eureka.instance.hostname}:${server.port}/eureka/
```

(3) 在启动类上加@EnableEurekaServer 注解。

```
@SpringBootApplication
@EnableEurekaServer
public class EurekaServerApplication {

    public static void main(String[] args) {
        SpringApplication.run(EurekaServerApplication.class, args);
    }

}
```

做完上面的步骤后，启动服务。

访问 http://localhost:8761/，如果出现如图 2-3 所示的界面（Eureka Server 管理控制台），说明 Eureka Server 启动成功。

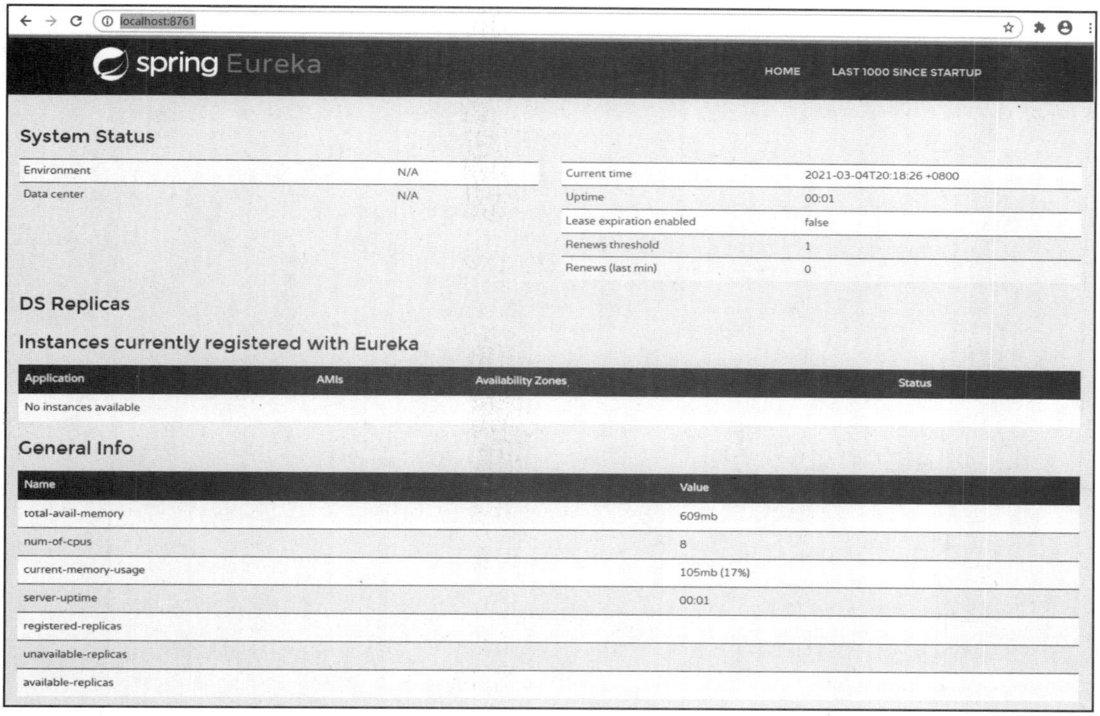

图 2-3　Eureka Server 启动界面

2.3.2 创建客户端

先创建一个 Spring Boot 项目 eureka-client，操作步骤如下。

（1）在 pom 文件中添加如下依赖。

```xml
<dependencies>
    <dependency>
        <groupId>org.springframework.boot</groupId>
        <artifactId>spring-boot-starter-web</artifactId>
    </dependency>

    <!--eureka client jar包-->
    <dependency>
        <groupId>org.springframework.cloud</groupId>
        <artifactId>spring-cloud-starter-netflix-eureka-client</artifactId>
    </dependency>

    <dependency>
        <groupId>org.springframework.boot</groupId>
        <artifactId>spring-boot-starter-test</artifactId>
        <scope>test</scope>
    </dependency>
</dependencies>
```

（2）配置 application.yml 文件。

```yml
server:
  port: 8080
spring:
  application:
    name: eureka-client
eureka:
  client:
    serviceUrl:
      # 服务注册中心地址
      defaultZone: http://localhost:8761/eureka/
```

做完上面的步骤后，启动服务。

观察 2.3.1 节中启动的 Eureka Server 界面，发现多了一个服务，如图 2-4 所示。

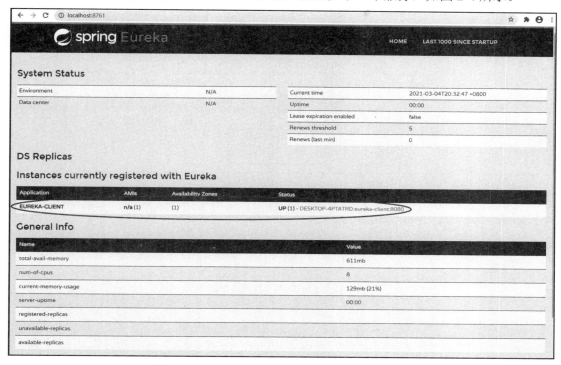

图 2-4 Eureka Client 注册到 Eureka Server 后的界面

2.3.3 Eureka Server 高可用搭建

服务之间在互相调用之前，会先去注册中心查一下需要调用的服务的地址，如果注册中心出现故障，那么整个系统中的服务调用就会全部失效，产生的后果是灾难性的。

为了避免上述故障，需要搭建高可用 Eureka Server。不同数量的 Eureka Server 集群的搭建方式是不一样的。

1. 两个 Eureka Server 的集群

为了大家看起来方便，使用域名需要配置一下本机的 hosts 文件，如图 2-5 所示。

现在创建一个项目 eureka-server-two-peer，只需要修改 application.yml 文件即可。我们创建两个 yml 文件，即 application-peer1.yml 和 application-peer2.yml。

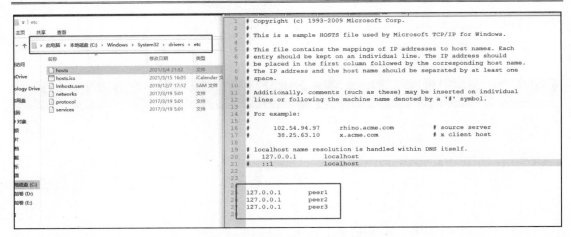

图 2-5　Windows 系统 hosts 文件配置

application-peer1.yml 文件如下所示。

```yaml
spring:
  application:
    name: eureka-server
server:
  port: 8761
eureka:
  instance:
    # 主机名
    hostname: peer1
  client:
    # 是否向 Eureka Server 注册，是否向自己注册
    registerWithEureka: true
    # 是否从 Eureka Server 拉取注册表
    fetchRegistry: true
    serviceUrl:
      # 对方 Eureka Server 的地址
      defaultZone: http://peer2:8762/eureka/
```

application-peer2.yml 文件如下所示。

```yaml
spring:
  application:
    name: eureka-server
```

```yaml
server:
  port: 8762
eureka:
  instance:
    # 主机名
    hostname: peer2
  client:
    # 是否向 Eureka Server 注册，是否向自己注册
    registerWithEureka: true
    # 是否从 Eureka Server 拉取注册表
    fetchRegistry: true
    serviceUrl:
      # 对方 Eureka Server 的地址
      defaultZone: http://peer1:8761/eureka/
```

下面在 IDE 中设置启动的 Active profiles，如图 2-6 和图 2-7 所示。

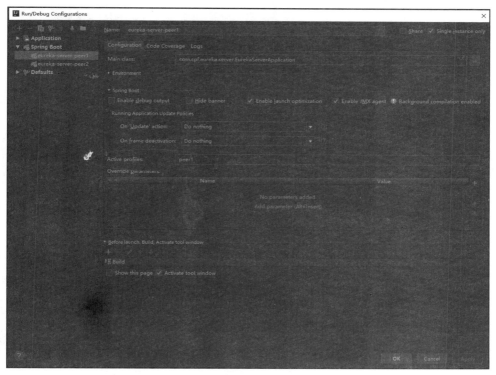

图 2-6　Eureka Server peer1 节点的 Active profiles 配置

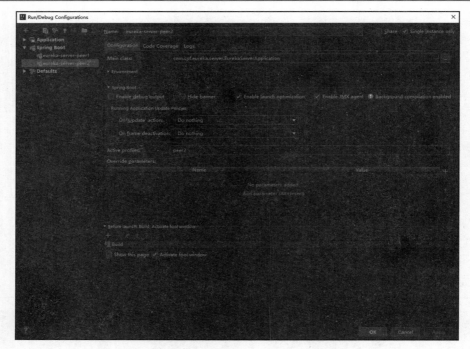

图 2-7　Eureka Server peer2 节点的 Active profiles 配置

启动上面配置的两个 Run/Debug Configurations，启动两个 Eureka Server。

访问 peer1，地址为 http://localhost:8761/，如图 2-8 所示。

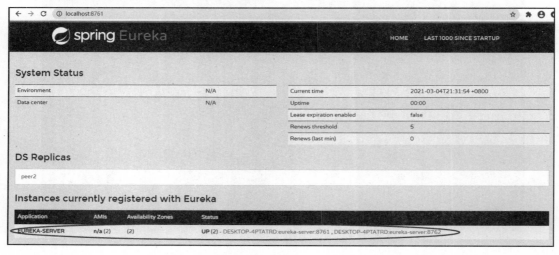

图 2-8　Eureka Server peer1 节点启动界面

访问 peer2，地址为 http://localhost:8762/，如图 2-9 所示。

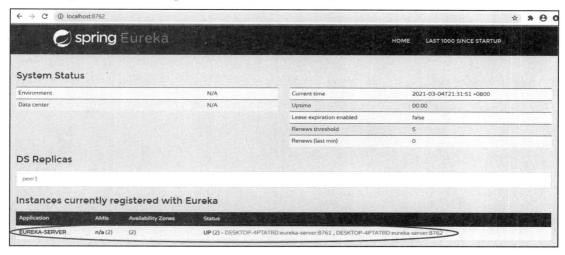

图 2-9　Eureka Server peer2 节点启动界面

如果能看到图 2-8 和图 2-9 所示的界面，就说明两个节点的 Eureka Server 集群搭建成功了。

2．三个 Eureka Server 节点的集群

三个节点的集群和两个节点类似，创建三个文件，即 application-peer1.yml、application-peer2.yml 和 application-peer3.yml。

application-peer1.yml 文件如下所示。

```yaml
spring:
  application:
    name: eureka-server
server:
  port: 8761
eureka:
  instance:
    # 主机名
    hostname: peer1
  client:
    # 是否向 Eureka Server 注册，是否向自己注册
    registerWithEureka: true
```

```
      # 是否从Eureka Server拉取注册表
      fetchRegistry: true
      serviceUrl:
        # 对方Eureka Server的地址
        defaultZone: http://peer1:8761/eureka/,http://peer2:8762/eureka/,http://peer3:8763/eureka/
```

application-peer2.yml文件如下所示。

```
spring:
  application:
    name: eureka-server
server:
  port: 8761
eureka:
  instance:
    # 主机名
    hostname: peer2
  client:
    # 是否向Eureka Server注册，是否向自己注册
    registerWithEureka: true
    # 是否从Eureka Server拉取注册表
    fetchRegistry: true
    serviceUrl:
      # 对方Eureka Server的地址
      defaultZone: http://peer1:8761/eureka/,http://peer2:8762/eureka/,http://peer3:8763/eureka/
```

application-peer3.yml文件如下所示。

```
spring:
  application:
    name: eureka-server
server:
  port: 8763
eureka:
  instance:
    # 主机名
    hostname: peer3
```

```
  client:
    # 是否向 Eureka Server 注册，是否向自己注册
    registerWithEureka: true
    # 是否从 Eureka Server 拉取注册表
    fetchRegistry: true
    serviceUrl:
      # 对方 Eureka Server 的地址
      defaultZone: http://peer1:8761/eureka/,http://peer2:8762/eureka/,http://peer3:8763/eureka/
```

启动后访问下面三个地址，启动界面如图 2-10 所示。

☑　peer1: http://localhost:8761/。
☑　peer2: http://localhost:8762/。
☑　peer3: http://localhost:8763/。

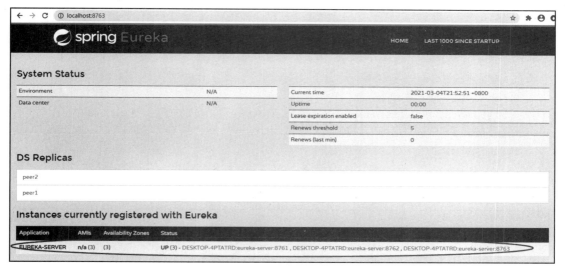

图 2-10　Eureka Server 每个 peer 节点启动界面

从上面的示例可以总结出，每个 Eureka Server 只需要在自己的配置文件中指定其他的 Eureka Server 地址就可以了。集群注册的原理其实很简单，如图 2-11 所示。

当 Eureka Client 向 Eureka Server1 注册时，Eureka Server1 会将注册信息同步给 Eureka Server2 和 Eureka Server3。同理，当 Eureka Client 向 Eureka Server2 注册时，Eureka Server2 也会将注册信息同步给 Eureka Server3 和 Eureka Server1，以此类推。

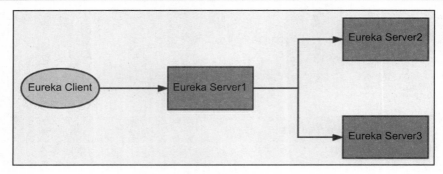

图 2-11 集群注册

2.3.4 Eureka Server 端用户认证

前面对 Eureka Server 端的操作，直接就能登录到控制台，在实际项目中这么做是不安全的。下面以单台 Eureka Server 节点来学习如何添加用户认证。先创建 eureka-server-security 项目，然后执行以下操作。

（1）添加以下依赖。

```xml
<!--添加 security jar包-->
<dependency>
    <groupId>org.springframework.boot</groupId>
    <artifactId>spring-boot-starter-security</artifactId>
</dependency>
```

（2）启动 Eureka Server。

观察控制台，可以看到一个密码，如图 2-12 所示。

```
2021-03-14 17:48:28.560 INFO 4524 --- [      main] c.n.d.provider.DiscoveryJerseyProvider   : Using JSON decoding codec LegacyJacksonJ
2021-03-14 17:48:28.677 INFO 4524 --- [      main] c.n.d.provider.DiscoveryJerseyProvider   : Using XML encoding codec XStreamXml
2021-03-14 17:48:28.677 INFO 4524 --- [      main] c.n.d.provider.DiscoveryJerseyProvider   : Using XML decoding codec XStreamXml
2021-03-14 17:48:29.218 INFO 4524 --- [      main] o.s.s.concurrent.ThreadPoolTaskExecutor  : Initializing ExecutorService 'applicatio
2021-03-14 17:48:30.512 INFO 4524 --- [      main] .s.s.UserDetailsServiceAutoConfiguration :

Using generated security password: 0e57354a-beb6-4acb-b92b-8817b68d5e48

2021-03-14 17:48:30.644 INFO 4524 --- [      main] o.s.s.web.DefaultSecurityFilterChain     : Will secure any request with [org.spring
2021-03-14 17:48:31.758 INFO 4524 --- [      main] DiscoveryClientOptionalArgsConfiguration : Eureka HTTP Client uses Jersey
2021-03-14 17:48:31.815 WARN 4524 --- [      main] iguration$LoadBalancerCaffeineWarnLogger : Spring Cloud LoadBalancer is currently w
2021-03-14 17:48:31.833 INFO 4524 --- [      main] o.s.c.n.eureka.InstanceInfoFactory       : Setting initial instance status as: STAR
2021-03-14 17:48:31.863 INFO 4524 --- [      main] com.netflix.discovery.DiscoveryClient    : Initializing Eureka in region us-east-1
```

图 2-12 Security 控制台密码

（3）进行用户认证。

Eureka Server 启动后，访问 http://localhost:8761/。在文本框中输入用户名 user，密码是

图 2-12 中控制台的那串字符串。输入后单击 Sign in 按钮就可以登录成功,如图 2-13 所示。

图 2-13 输入用户名和密码

(4)自定义用户名和密码。

上面的操作有一个问题,就是每次启动 Eureka Server 时,密码是动态变化的,这样不利于管理。因此需要自定义用户名和密码。方法也简单,修改 application.yml 文件即可。

```yaml
spring:
  security:
    user:
      # 用户名
      name: root
      # 密码
      password: 123456
```

这样,用户名、密码就分别改成了 root、123456。

(5)改造 Eureka Client。

在 Eureka Server 端加了验证后,启动 Eureka Client 时会报错。因为 Eureka Client 没有填写用户名和密码,所以注册不上 Eureka Server。在 Eureka Client 端修改 application.yml。

```yaml
eureka:
  client:
    serviceUrl:
      # 在注册中心地址中,配置用户名和密码
      defaultZone: http://root:123456@localhost:8761/eureka/
```

同时,在 Spring Cloud 2.0 后,Spring Security 默认开启了 CSRF 校验,所以必须在 Eureka Server 端禁用/eureka/**的 CSRF 校验。在 Eureka Server 端添加如下代码。

```java
@EnableWebSecurity
public class WebSecurityConfig extends WebSecurityConfigurerAdapter {
```

```java
@Override
protected void configure(HttpSecurity http) throws Exception {
    http.csrf().ignoringAntMatchers("/eureka/**");
    super.configure(http);
}
}
```

此时重启 Eureka Server 和 Eureka Client，即可注册成功。

2.3.5　自我保护机制

客户端会通过定期向注册中心服务端发送心跳，来让注册中心的服务端知道客户端可用，如果没有心跳，Eureka Server 会将其从注册表中删除。在 Eureka Server 中有一项自我保护机制，它默认是开启状态。在 EurekaServerConfigBean 类源码中，有下面一行代码。

```
# 自我保护机制默认开启
private boolean enableSelfPreservation = true;
```

自我保护机制开启时，有时候会在 Eureka Server 的管理界面看到一行提示文字，如图 2-14 所示。

图 2-14　自我保护提示界面

开启自我保护，是因为在微服务环境中，有时由于网络抖动，正常的服务没有向 Eureka Server 发送心跳，Eureka Server 会有一个统计比例，默认是 85%。当心跳的比例低于应该收到的正常心跳的 85% 时，Eureka Server 会将这些服务保护起来，不剔除。Eureka Server 此时宁可将已经不可用的服务保留，也不会因为网络抖动剔除正常可用的服务。

那么当服务调用到上面保留的不可用服务时，应该怎么办呢？这个问题可通过熔断方案来解决，后面内容会讲。

如果想关闭自我保护，在 application.yml 中修改配置即可。

```
eureka:
  server:
    # 关闭注册中心
    enable-self-preservation: false
```

大厂面试

面试官：自我保护机制开启后，是不是就不进行服务的剔除了？

面试者：不一定。自我保护开启后，剔除机制还需要进行判断，判断最近一分钟的续约数是否大于配置的阈值，如果大于阈值，且服务达到剔除条件，正常进行剔除；如果小于阈值，不剔除（此处是最终的自我保护生效的地方）。

分析：可以向面试官画一个图来进行解释，如图 2-15 所示。

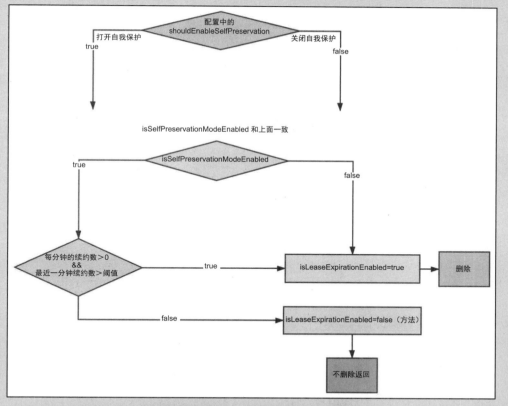

图 2-15　自我保护流程

2.3.6 多网卡选择

在生产环境中，有时候一台机器配有多个网卡，所以就有多个 IP 地址。例如，一台机器有外网 IP，有内外 IP，而内网的 IP 地址是无法被外网的其他服务访问的。如果服务将不可被其他服务访问的 IP 注册到注册中心，就会导致以后调用不通。所以需要在 application.yml 中修改注册的 IP 地址。

```
eureka:
  instance:
    # 配置用 IP
    prefer-ip-address: true
    # 配置具体的 IP 地址
    ip-address: 127.0.0.1
```

修改配置后，指定 IP 注册后的管理界面如图 2-16 所示。

图 2-16 指定 IP 注册后的管理界面

2.3.7 Eureka Server 源码解析

本节从 Eureka Server 的源码角度来进行 Eureka 的学习。这里要先明确一点，Eureka Server 也是一个 Eureka Client，在不禁止 Eureka Server 的客户端行为时，它会向配置文件中所配置的 Eureka Server 进行拉取注册表、服务注册和发送心跳的操作。

Eureka Server 提供了以下几个功能。

1. 接收服务注册

接收服务注册其实也是一个 Web 服务。Eureka Server 通过 ApplicationResource 提供了 Web

接口。注册服务的入口代码如下。

```java
@POST
@Consumes({"application/json", "application/xml"})
public Response addInstance(InstanceInfo info, @HeaderParam("x-netflix-discovery-replication") String isReplication) {
    logger.debug("Registering instance {} (replication={})", info.getId(), isReplication);
    // 执行注册的参数校验
    if (this.isBlank(info.getId())) {
        return Response.status(400).entity("Missing instanceId").build();
    } else if (this.isBlank(info.getHostName())) {
        return Response.status(400).entity("Missing hostname").build();
    } else if (this.isBlank(info.getIPAddr())) {
        return Response.status(400).entity("Missing ip address").build();
    } else if (this.isBlank(info.getAppName())) {
        return Response.status(400).entity("Missing appName").build();
    } else if (!this.appName.equals(info.getAppName())) {
        return Response.status(400).entity("Mismatched appName, expecting " + this.appName + " but was " + info.getAppName()).build();
    } else if (info.getDataCenterInfo() == null) {
        return Response.status(400).entity("Missing dataCenterInfo").build();
    } else if (info.getDataCenterInfo().getName() == null) {
        return Response.status(400).entity("Missing dataCenterInfo Name").build();
    } else {
        DataCenterInfo dataCenterInfo = info.getDataCenterInfo();
        if (dataCenterInfo instanceof UniqueIdentifier) {
            String dataCenterInfoId = ((UniqueIdentifier)dataCenterInfo).getId();
            if (this.isBlank(dataCenterInfoId)) {
                boolean experimental = "true".equalsIgnoreCase(this.serverConfig.getExperimental("registration.validation.dataCenterInfoId"));
                if (experimental) {
                    String entity = "DataCenterInfo of type " + dataCenterInfo.getClass() + " must contain a valid id";
                    return Response.status(400).entity(entity).build();
```

```
                    }
                    if (dataCenterInfo instanceof AmazonInfo) {
                        AmazonInfo amazonInfo = (AmazonInfo)dataCenterInfo;
                        String effectiveId = amazonInfo.get(MetaDataKey.
instanceId);
                        if (effectiveId == null) {
                            amazonInfo.getMetadata().put(MetaDataKey.
instanceId.getName(), info.getId());
                        }
                    } else {
                        logger.warn("Registering DataCenterInfo of type {}
without an appropriate id", dataCenterInfo.getClass());
                    }
                }
            }
            // 执行实际的注册逻辑
            this.registry.register(info, "true".equals(isReplication));
            return Response.status(204).build();
        }
    }
```

打开 InstanceRegistry 类，InstanceRegistry 类是 Eureka 管理注册表的核心类。查看以下方法。

```
@Override
public void register(final InstanceInfo info, final boolean isReplication) {
handleRegistration(info, resolveInstanceLeaseDuration(info), isReplication);
super.register(info, isReplication);
}
```

此方法接收 Eureka Client 发过来的服务实例信息，并将信息保存到注册表中。其中的几个参数说明如下。

- ☑ InstanceInfo：服务实例的信息，包括实例 ID、应用名称、IP 地址、端口、元数据等和服务实例相关的信息。
- ☑ resolveInstanceLeaseDuration：租约有效期，即当前服务实例的有效期，过了这个有效期，如果没有心跳就剔除。

- isReplication：是否同步，收到服务实例来注册，是否要将信息同步到其他 Eureka Server 节点。

以上代码中还调用了 handleRegistration 方法，这个方法主要是进行注册事件 EurekaInstanceRegisteredEvent 的发布，这样可以基于 EurekaInstanceRegisteredEvent 事件进行后续的逻辑处理。

```
private void handleRegistration(InstanceInfo info, int leaseDuration,
boolean isReplication) {
log("register " + info.getAppName() + ", vip " + info.getVIPAddress() + ",
leaseDuration " + leaseDuration + ", isReplication " + isReplication);
publishEvent(new EurekaInstanceRegisteredEvent(this, info, leaseDuration,
isReplication));
}
```

从这里得知，可以通过对事件 EurekaInstanceRegisteredEvent 的监听，来自定义收到服务注册后的处理逻辑。

在 handleRegistration 方法执行后，执行了父类的注册方法，代码如下。

```
public void register(InstanceInfo registrant, int leaseDuration, boolean isReplication) {
    // 获取读锁
    this.read.lock();
    
    try {
        // 根据服务名获取原来的注册表，就像前面的分析，注册表用的是 Map 进行存储
        Map<String, Lease<InstanceInfo>> gMap = (Map)this.registry.get
(registrant.getAppName());
        EurekaMonitors.REGISTER.increment(isReplication);
        // 如果原来没有注册表，则新建一个 Map
        if (gMap == null) {
            ConcurrentHashMap<String, Lease<InstanceInfo>> gNewMap = new
ConcurrentHashMap();
            // 这个操作，防止在新添加服务实例注册信息时，将老的服务实例覆盖。此时如果原来已经有该键，则返回原来已经存在的值，否则添加该键-值对
            gMap = (Map)this.registry.putIfAbsent(registrant.
getAppName(), gNewMap);
            if (gMap == null) {
                gMap = gNewMap;
```

```java
            }
        }
        // 根据服务实例ID，获取租约信息
        Lease<InstanceInfo> existingLease = (Lease)((Map)gMap).get(registrant.getId());
        if (existingLease != null && existingLease.getHolder() != null) {
            Long existingLastDirtyTimestamp = ((InstanceInfo)existingLease.getHolder()).getLastDirtyTimestamp();
            Long registrationLastDirtyTimestamp = registrant.getLastDirtyTimestamp();
            logger.debug("Existing lease found (existing={}, provided={}", existingLastDirtyTimestamp, registrationLastDirtyTimestamp);
            if (existingLastDirtyTimestamp > registrationLastDirtyTimestamp) {
                logger.warn("There is an existing lease and the existing lease's dirty timestamp {} is greater than the one that is being registered {}", existingLastDirtyTimestamp, registrationLastDirtyTimestamp);
                logger.warn("Using the existing instanceInfo instead of the new instanceInfo as the registrant");
                registrant = (InstanceInfo)existingLease.getHolder();
            }
        } else {
            Object var6 = this.lock;
            synchronized(this.lock) {
                // 设置自我保护机制的参数
                if (this.expectedNumberOfClientsSendingRenews > 0) {
                    ++this.expectedNumberOfClientsSendingRenews;
                    this.updateRenewsPerMinThreshold();
                }
            }

            logger.debug("No previous lease information found; it is new registration");
        }
        // 创建新的租约
        Lease<InstanceInfo> lease = new Lease(registrant, leaseDuration);
        if (existingLease != null) {
```

```java
            lease.setServiceUpTimestamp(existingLease.getServiceUpTimestamp());
        }
        // 保存租约
        ((Map)gMap).put(registrant.getId(), lease);
        // 添加最近注册队列
        this.recentRegisteredQueue.add(new Pair(System.currentTimeMillis(), registrant.getAppName() + "(" + registrant.getId() + ")"));
        if (!InstanceStatus.UNKNOWN.equals(registrant.getOverriddenStatus())) {
            logger.debug("Found overridden status {} for instance {}. Checking to see if needs to be add to the overrides", registrant.getOverriddenStatus(), registrant.getId());
            if (!this.overriddenInstanceStatusMap.containsKey(registrant.getId())) {
                logger.info("Not found overridden id {} and hence adding it", registrant.getId());
                this.overriddenInstanceStatusMap.put(registrant.getId(), registrant.getOverriddenStatus());
            }
        }

        InstanceStatus overriddenStatusFromMap = (InstanceStatus)this.overriddenInstanceStatusMap.get(registrant.getId());
        if (overriddenStatusFromMap != null) {
            logger.info("Storing overridden status {} from map", overriddenStatusFromMap);
            registrant.setOverriddenStatus(overriddenStatusFromMap);
        }

        InstanceStatus overriddenInstanceStatus = this.getOverriddenInstanceStatus(registrant, existingLease, isReplication);
        registrant.setStatusWithoutDirty(overriddenInstanceStatus);
        if (InstanceStatus.UP.equals(registrant.getStatus())) {
            lease.serviceUp();
        }
```

```
        // 设置此时的动作为服务实例的新增
        registrant.setActionType(ActionType.ADDED);
        this.recentlyChangedQueue.add(new AbstractInstanceRegistry.
RecentlyChangedItem(lease));
        // 设置服务实例信息更新时间
        registrant.setLastUpdatedTimestamp();
        this.invalidateCache(registrant.getAppName(), registrant.
getVIPAddress(), registrant.getSecureVipAddress());
        logger.info("Registered instance {}/{} with status {} 
(replication={})", new Object[]{registrant.getAppName(), registrant.
getId(), registrant.getStatus(), isReplication});
    } finally {
        this.read.unlock();
    }
}
```

2．接收服务心跳

心跳入口的 Web 接口代码如下所示。

```
@PUT
public Response renewLease(@HeaderParam("x-netflix-discovery-replication") 
String isReplication, @QueryParam("overriddenstatus") String 
overriddenStatus, @QueryParam("status") String status, @QueryParam
("lastDirtyTimestamp") String lastDirtyTimestamp) {
        boolean isFromReplicaNode = "true".equals(isReplication);
        boolean isSuccess = this.registry.renew(this.app.getName(), this.id, 
isFromReplicaNode);
        if (!isSuccess) {
            logger.warn("Not Found (Renew): {} - {}", this.app.getName(), 
this.id);
            return Response.status(Status.NOT_FOUND).build();
        } else {
            Response response;
            if (lastDirtyTimestamp != null && this.serverConfig.
shouldSyncWhenTimestampDiffers()) {
                response = this.validateDirtyTimestamp(Long.valueOf
(lastDirtyTimestamp), isFromReplicaNode);
```

```
            if (response.getStatus() == Status.NOT_FOUND.getStatusCode()
&& overriddenStatus != null && !InstanceStatus.UNKNOWN.name().equals
(overriddenStatus) && isFromReplicaNode) {
                this.registry.storeOverriddenStatusIfRequired(this.app.
getAppName(), this.id, InstanceStatus.valueOf(overriddenStatus));
            }
        } else {
            response = Response.ok().build();
        }

        logger.debug("Found (Renew): {} - {}; reply status={}", new
Object[]{this.app.getName(), this.id, response.getStatus()});
        return response;
    }
}
```

在 InstanceRegistry.java 类中有一个 renew 方法。

```
@Override
public boolean renew(final String appName, final String serverId, boolean
isReplication) {
    log("renew " + appName + " serverId " + serverId + ", isReplication {}"
+ isReplication);
    List<Application> applications = getSortedApplications();
    for (Application input : applications) {
        if (input.getName().equals(appName)) {
            InstanceInfo instance = null;
            for (InstanceInfo info : input.getInstances()) {
                if (info.getId().equals(serverId)) {
                    instance = info;
                    break;
                }
            }
            // 发布心跳事件
            publishEvent(new EurekaInstanceRenewedEvent(this, appName,
serverId, instance, isReplication));
            break;
        }
```

```
    }
    // 执行心跳业务逻辑
    return super.renew(appName, serverId, isReplication);
}
```

在 PeerAwareInstanceRegistryImpl.java 类中也有一个 renew 方法。

```
public boolean renew(String appName, String id, boolean isReplication) {
    if (super.renew(appName, id, isReplication)) {
        // 将心跳同步到集群中的其他 Server 节点
        this.replicateToPeers(PeerAwareInstanceRegistryImpl.Action.Heartbeat, appName, id, (InstanceInfo)null, (InstanceStatus)null, isReplication);
        return true;
    } else {
        return false;
    }
}
```

3. 接收服务下线

服务下线的入口代码如下所示。

```
@DELETE
public Response cancelLease(@HeaderParam("x-netflix-discovery-replication")
String isReplication) {
    try {
        boolean isSuccess = this.registry.cancel(this.app.getName(), this.id, "true".equals(isReplication));
        if (isSuccess) {
            logger.debug("Found (Cancel): {} - {}", this.app.getName(), this.id);
            return Response.ok().build();
        } else {
            logger.info("Not Found (Cancel): {} - {}", this.app.getName(), this.id);
            return Response.status(Status.NOT_FOUND).build();
        }
    } catch (Throwable var3) {
```

```
        logger.error("Error (cancel): {} - {}", new Object[]{this.app.
getName(), this.id, var3});
        return Response.serverError().build();
    }
}
```

在 InstanceRegistry 类中有一个 cancel 方法。

```
@Override
public boolean cancel(String appName, String serverId, boolean isReplication) {
    handleCancelation(appName, serverId, isReplication);
    return super.cancel(appName, serverId, isReplication);
}
```

调用 handleCancelation 方法进行服务实例取消事件的发布。

```
private void handleCancelation(String appName, String id, boolean isReplication) {
    log("cancel " + appName + ", serverId " + id + ", isReplication " + isReplication);
    publishEvent(new EurekaInstanceCanceledEvent(this, appName, id, isReplication));
}
```

调用 super.cancel 方法进行服务实例的取消。

```
public boolean cancel(String appName, String id, boolean isReplication) {
    // 执行取消的具体操作
    if (super.cancel(appName, id, isReplication)) {
        // 同步到其他 Eureka Server 节点
        this.replicateToPeers(PeerAwareInstanceRegistryImpl.Action.Cancel, appName, id, (InstanceInfo)null, (InstanceStatus)null, isReplication);
        return true;
    } else {
        return false;
    }
}
```

最终取消的方法代码如下所示。

```
protected boolean internalCancel(String appName, String id, boolean isReplication) {
```

```java
        // 获取读锁
        this.read.lock();

        label97: {
            boolean var7;
            try {
                EurekaMonitors.CANCEL.increment(isReplication);
                Map<String, Lease<InstanceInfo>> gMap = (Map)this.registry.get(appName);
                Lease<InstanceInfo> leaseToCancel = null;
                if (gMap != null) {
                    // 移除服务实例
                    leaseToCancel = (Lease)gMap.remove(id);
                }
                // 在最近取消服务实例的队列中，添加信息
                this.recentCanceledQueue.add(new Pair(System.currentTimeMillis(), appName + "(" + id + ")"));
                InstanceStatus instanceStatus = (InstanceStatus)this.overriddenInstanceStatusMap.remove(id);
                if (instanceStatus != null) {
                    logger.debug("Removed instance id {} from the overridden map which has value {}", id, instanceStatus.name());
                }

                if (leaseToCancel != null) {
                    leaseToCancel.cancel();
                    InstanceInfo instanceInfo = (InstanceInfo)leaseToCancel.getHolder();
                    String vip = null;
                    String svip = null;
                    if (instanceInfo != null) {
                        // 添加动作为删除服务实例
                        instanceInfo.setActionType(ActionType.DELETED);
                        this.recentlyChangedQueue.add(new AbstractInstanceRegistry.RecentlyChangedItem(leaseToCancel));
                        instanceInfo.setLastUpdatedTimestamp();
                        vip = instanceInfo.getVIPAddress();
```

```
                svip = instanceInfo.getSecureVipAddress();
            }

            this.invalidateCache(appName, vip, svip);
            logger.info("Cancelled instance {}/{} (replication={})",
new Object[]{appName, id, isReplication});
            break label97;
        }

        EurekaMonitors.CANCEL_NOT_FOUND.increment(isReplication);
        logger.warn("DS: Registry: cancel failed because Lease is not
registered for: {}/{}", appName, id);
        var7 = false;
    } finally {
        this.read.unlock();
    }

    return var7;
}
```

4. 服务剔除

在 AbstractInstanceRegistry 中有一个 postInit 方法。

```
protected void postInit() {
    this.renewsLastMin.start();
    if (this.evictionTaskRef.get() != null) {
        ((AbstractInstanceRegistry.EvictionTask)this.evictionTaskRef.get()).
cancel();
    }

    this.evictionTaskRef.set(new AbstractInstanceRegistry.EvictionTask());
    this.evictionTimer.schedule((TimerTask)this.evictionTaskRef.get(),
this.serverConfig.getEvictionIntervalTimerInMs(), this.serverConfig.
getEvictionIntervalTimerInMs());
}
```

这里重要的是一个定时任务 EvictionTask，实际剔除执行的逻辑代码如下所示。

```java
public void evict(long additionalLeaseMs) {
    logger.debug("Running the evict task");
    if (!this.isLeaseExpirationEnabled()) {
        logger.debug("DS: lease expiration is currently disabled.");
    } else {
        List<Lease<InstanceInfo>> expiredLeases = new ArrayList();
        Iterator var4 = this.registry.entrySet().iterator();

        while(true) {
            Map leaseMap;
            do {
                if (!var4.hasNext()) {
                    int registrySize = (int)this.getLocalRegistrySize();
                    // 计算剔除的阈值
                    int registrySizeThreshold = (int)((double)registrySize * this.serverConfig.getRenewalPercentThreshold());
                    // 计算要剔除服务的数量
                    int evictionLimit = registrySize - registrySizeThreshold;
                    int toEvict = Math.min(expiredLeases.size(), evictionLimit);
                    if (toEvict > 0) {
                        logger.info("Evicting {} items (expired={}, evictionLimit={})",new Object[]{toEvict,expiredLeases.size(),evictionLimit});
                        Random random = new Random(System.currentTimeMillis());

                        for(int i = 0; i < toEvict; ++i) {
                            int next = i + random.nextInt(expiredLeases.size() - i);
                            Collections.swap(expiredLeases, i, next);
                            Lease<InstanceInfo> lease = (Lease)expiredLeases.get(i);
                            String appName = ((InstanceInfo)lease.getHolder()).getAppName();
                            String id=((InstanceInfo)lease.getHolder()).getId();
                            EurekaMonitors.EXPIRED.increment();
                            logger.warn("DS: Registry: expired lease for {}/{}",appName, id);
                            // 剔除时，实际执行的是取消方法（前面讲过）
```

```
                this.internalCancel(appName, id, false);
            }
        }
        return;
    }

        Entry<String, Map<String, Lease<InstanceInfo>>> groupEntry = (Entry)var4.next();
        leaseMap = (Map)groupEntry.getValue();
    } while(leaseMap == null);

    Iterator var7 = leaseMap.entrySet().iterator();

    while(var7.hasNext()) {
        Entry<String, Lease<InstanceInfo>> leaseEntry = (Entry)var7.next();
        Lease<InstanceInfo> lease = (Lease)leaseEntry.getValue();
        if (lease.isExpired(additionalLeaseMs) && lease.getHolder() != null) {
            expiredLeases.add(lease);
        }
    }
  }
 }
}
```

5. 查询注册表

查询注册表的入口, 在 ApplicationResource 的 getApplication 方法中。

```
@GET
public Response getApplication(@PathParam("version") String version,
@HeaderParam("Accept") String acceptHeader, @HeaderParam("X-Eureka-Accept")
String eurekaAccept) {
    if (!this.registry.shouldAllowAccess(false)) {
        return Response.status(Status.FORBIDDEN).build();
    } else {
        EurekaMonitors.GET_APPLICATION.increment();
        CurrentRequestVersion.set(Version.toEnum(version));
```

```java
            KeyType keyType = KeyType.JSON;
            if (acceptHeader == null || !acceptHeader.contains("json")) {
                keyType = KeyType.XML;
            }

            Key cacheKey = new Key(EntityType.Application, this.appName, keyType,
CurrentRequestVersion.get(), EurekaAccept.fromString(eurekaAccept));
            String payLoad = this.responseCache.get(cacheKey);
            CurrentRequestVersion.remove();
            if (payLoad != null) {
                logger.debug("Found: {}", this.appName);
                return Response.ok(payLoad).build();
            } else {
                logger.debug("Not Found: {}", this.appName);
                return Response.status(Status.NOT_FOUND).build();
            }
        }
    }
}
```

最终读取注册表的代码如下所示。

```java
@VisibleForTesting
ResponseCacheImpl.Value getValue(Key key, boolean useReadOnlyCache) {
    ResponseCacheImpl.Value payload = null;
    try {
        if (useReadOnlyCache) {
            ResponseCacheImpl.Value currentPayload = (ResponseCacheImpl.
Value)this.readOnlyCacheMap.get(key);
            if (currentPayload != null) {
                payload = currentPayload;
            } else {
                payload = (ResponseCacheImpl.Value)this.readWriteCacheMap.
get(key);
                this.readOnlyCacheMap.put(key, payload);
            }
        } else {
            payload = (ResponseCacheImpl.Value)this.readWriteCacheMap.
get(key);
```

```
        }
    } catch (Throwable var5) {
        logger.error("Cannot get value for key : {}", key, var5);
    }
    return payload;
}
```

ResponseCacheImpl 初始的时候,是用 Guava 做的缓存组件。

6. 集群间注册表同步

集群间同步的核心是 PeerAwareInstanceRegistryImpl 类中的 replicateToPeers 方法。

```
private void replicateToPeers(PeerAwareInstanceRegistryImpl.Action action,
String appName, String id, InstanceInfo info, InstanceStatus newStatus,
boolean isReplication) {
    Stopwatch tracer = action.getTimer().start();
    try {
        if (isReplication) {
            this.numberOfReplicationsLastMin.increment();
        }
        if (this.peerEurekaNodes != Collections.EMPTY_LIST && !isReplication) {
            Iterator var8 = this.peerEurekaNodes.getPeerEurekaNodes().
iterator();
            while(var8.hasNext()) {
                PeerEurekaNode node = (PeerEurekaNode)var8.next();
                if (!this.peerEurekaNodes.isThisMyUrl(node.getServiceUrl())) {
                    this.replicateInstanceActionsToPeers(action, appName, id,
info, newStatus, node);
                }
            }
            return;
        }
    } finally {
        tracer.stop();
    }
}
```

下面进入 replicateInstanceActionsToPeers 方法。

```java
private void replicateInstanceActionsToPeers(PeerAwareInstanceRegistryImpl.Action action, String appName, String id, InstanceInfo info, InstanceStatus newStatus, PeerEurekaNode node) {
    try {
        CurrentRequestVersion.set(Version.V2);
        InstanceInfo infoFromRegistry;
        switch(action) {
        // 服务取消事件
        case Cancel:
            node.cancel(appName, id);
            break;
        // 服务心跳事件
        case Heartbeat:
            InstanceStatus overriddenStatus = (InstanceStatus)this.overriddenInstanceStatusMap.get(id);
            infoFromRegistry = this.getInstanceByAppAndId(appName, id, false);
            node.heartbeat(appName, id, infoFromRegistry, overriddenStatus, false);
            break;
        // 服务注册事件
        case Register:
            node.register(info);
            break;
        // 服务状态改变事件
        case StatusUpdate:
            infoFromRegistry = this.getInstanceByAppAndId(appName, id, false);
            node.statusUpdate(appName, id, newStatus, infoFromRegistry);
            break;
        case DeleteStatusOverride:
            infoFromRegistry = this.getInstanceByAppAndId(appName, id, false);
            node.deleteStatusOverride(appName, id, infoFromRegistry);
        }
    } catch (Throwable var12) {
        logger.error("Cannot replicate information to {} for action {}", new Object[]{node.getServiceUrl(), action.name(), var12});
    } finally {
```

```
        CurrentRequestVersion.remove();
    }

}
```

2.3.8　Eureka Client 源码解析

和 Eureka Server 相对应，Eureka Client 的主要操作就是调用 Eureka Server 提供的接口，其主要的类是 com.netflix.discovery.DiscoveryClient，在它的构造函数中实现了 Eureka Client 的几乎所有与 Eureka Server 相关的功能。

在构造函数初始化的过程中，执行了以下任务。

- ☑ 读取与 Eureka Server 交互的配置信息，封装成 EurekaClientConfig。
- ☑ 读取自身配置信息，封装成 EurekaInstanceConfig。
- ☑ 从 Eureka Server 端拉取注册信息，缓存到本地。
- ☑ 进行服务注册。
- ☑ 初始化 3 个定时器。
- ☑ 定时发送心跳到 Eureka Server 端，维持在注册表中的服务租约。
- ☑ 定时从 Eureka Server 端拉取注册表信息，更新本地缓存。
- ☑ 监控自身变化，如果有变化了，则重新发起注册。
- ☑ 从 Eureka Server 端销毁自身。一般情况下，应用服务在关闭时，Eureka Client 会主动向 Eureka Server 注销自身在注册表中的信息。

下面来看一下构造函数的代码。

```
@Inject
DiscoveryClient(ApplicationInfoManager applicationInfoManager,
EurekaClientConfig config, AbstractDiscoveryClientOptionalArgs args,
Provider<BackupRegistry> backupRegistryProvider, EndpointRandomizer
endpointRandomizer) {
    if (args != null) {
        this.healthCheckHandlerProvider = args.healthCheckHandlerProvider;
        this.healthCheckCallbackProvider = args.healthCheckCallbackProvider;
        this.eventListeners.addAll(args.getEventListeners());
        this.preRegistrationHandler = args.preRegistrationHandler;
    } else {
        this.healthCheckCallbackProvider = null;
```

```java
        this.healthCheckHandlerProvider = null;
        this.preRegistrationHandler = null;
    }

    this.applicationInfoManager = applicationInfoManager;
    InstanceInfo myInfo = applicationInfoManager.getInfo();

    clientConfig = config;
    staticClientConfig = clientConfig;
    transportConfig = config.getTransportConfig();
    instanceInfo = myInfo;
    if (myInfo != null) {
        appPathIdentifier = instanceInfo.getAppName() + "/" + instanceInfo.getId();
    } else {
        logger.warn("Setting instanceInfo to a passed in null value");
    }

    this.backupRegistryProvider = backupRegistryProvider;
    this.endpointRandomizer = endpointRandomizer;
    this.urlRandomizer = new EndpointUtils.InstanceInfoBasedUrlRandomizer(instanceInfo);
    localRegionApps.set(new Applications());

    fetchRegistryGeneration = new AtomicLong(0);

    remoteRegionsToFetch = new AtomicReference<String>(clientConfig.fetchRegistryForRemoteRegions());
    remoteRegionsRef = new AtomicReference<>(remoteRegionsToFetch.get() == null ? null : remoteRegionsToFetch.get().split(","));
    // 是否去注册中心拉取注册表,这里是通过在 yml 中的配置来指定的
    if (config.shouldFetchRegistry()) {
        this.registryStalenessMonitor = new ThresholdLevelsMetric(this, METRIC_REGISTRY_PREFIX + "lastUpdateSec_", new long[]{15L, 30L, 60L, 120L, 240L, 480L});
    } else {
        this.registryStalenessMonitor = ThresholdLevelsMetric.NO_OP_METRIC;
    }
```

```java
    // 判断是否向注册中心注册
    if (config.shouldRegisterWithEureka()) {
        this.heartbeatStalenessMonitor = new ThresholdLevelsMetric(this,
METRIC_REGISTRATION_PREFIX + "lastHeartbeatSec_", new long[]{15L, 30L, 60L,
120L, 240L, 480L});
    } else {
        this.heartbeatStalenessMonitor = ThresholdLevelsMetric.NO_OP_METRIC;
    }

    logger.info("Initializing Eureka in region {}", clientConfig.
getRegion());

    if (!config.shouldRegisterWithEureka() && !config.shouldFetchRegistry()) {
        logger.info("Client configured to neither register nor query for
data.");
        scheduler = null;
        heartbeatExecutor = null;
        cacheRefreshExecutor = null;
        eurekaTransport = null;
        instanceRegionChecker = new InstanceRegionChecker(new
PropertyBasedAzToRegionMapper(config), clientConfig.getRegion());

        // 使用 DiscoveryManager.getInstance()获取实例
        // 使用 DiscoveryClient
        DiscoveryManager.getInstance().setDiscoveryClient(this);
        DiscoveryManager.getInstance().setEurekaClientConfig(config);

        initTimestampMs = System.currentTimeMillis();
        initRegistrySize = this.getApplications().size();
        registrySize = initRegistrySize;
        logger.info("Discovery Client initialized at timestamp {} with
initial instances count: {}", initTimestampMs, initRegistrySize);

        return;   // 无须创建网络任务，方法结束
    }

    try {
        // 调度线程池默认大小为 2，心跳和缓存刷新各有一个线程池
```

```java
        // 创建一个Schedule线程池
        scheduler = Executors.newScheduledThreadPool(2,
                new ThreadFactoryBuilder()
                        .setNameFormat("DiscoveryClient-%d")
                        .setDaemon(true)
                        .build());
        // 心跳线程
        heartbeatExecutor = new ThreadPoolExecutor(
                1, clientConfig.getHeartbeatExecutorThreadPoolSize(), 0, TimeUnit.SECONDS,
                new SynchronousQueue<Runnable>(),
                new ThreadFactoryBuilder()
                        .setNameFormat("DiscoveryClient-HeartbeatExecutor-%d")
                        .setDaemon(true)
                        .build()
        );  // 使用直接切换
        // 缓存刷新线程
        cacheRefreshExecutor = new ThreadPoolExecutor(
                1, clientConfig.getCacheRefreshExecutorThreadPoolSize(), 0, TimeUnit.SECONDS,
                new SynchronousQueue<Runnable>(),
                new ThreadFactoryBuilder()
                        .setNameFormat("DiscoveryClient-CacheRefreshExecutor-%d")
                        .setDaemon(true)
                        .build()
        );  // 使用直接切换

        eurekaTransport = new EurekaTransport();
        scheduleServerEndpointTask(eurekaTransport, args);

        AzToRegionMapper azToRegionMapper;
        if (clientConfig.shouldUseDnsForFetchingServiceUrls()) {
            azToRegionMapper = new DNSBasedAzToRegionMapper(clientConfig);
        } else {
            azToRegionMapper = new PropertyBasedAzToRegionMapper(clientConfig);
        }
```

```java
        if (null != remoteRegionsToFetch.get()) {
            azToRegionMapper.setRegionsToFetch(remoteRegionsToFetch.get().split(","));
        }
        instanceRegionChecker = new InstanceRegionChecker(azToRegionMapper, clientConfig.getRegion());
    } catch (Throwable e) {
        throw new RuntimeException("Failed to initialize DiscoveryClient!", e);
    }

    if (clientConfig.shouldFetchRegistry()) {
        try {
            // 拉取注册表
            boolean primaryFetchRegistryResult = fetchRegistry(false);
            if (!primaryFetchRegistryResult) {
                logger.info("Initial registry fetch from primary servers failed");
            }
            boolean backupFetchRegistryResult = true;
            if (!primaryFetchRegistryResult && !fetchRegistryFromBackup()) {
                backupFetchRegistryResult = false;
                logger.info("Initial registry fetch from backup servers failed");
            }
            if (!primaryFetchRegistryResult && !backupFetchRegistryResult && clientConfig.shouldEnforceFetchRegistryAtInit()) {
                throw new IllegalStateException("Fetch registry error at startup. Initial fetch failed.");
            }
        } catch (Throwable th) {
            logger.error("Fetch registry error at startup: {}", th.getMessage());
            throw new IllegalStateException(th);
        }
    }

    // 在所有背景任务（如注册）启动前，调用并执行预注册方法
    if (this.preRegistrationHandler != null) {
```

```java
            this.preRegistrationHandler.beforeRegistration();
        }

        if (clientConfig.shouldRegisterWithEureka() && clientConfig.shouldEnforceRegistrationAtInit()) {
            try {
                // 这个判断中执行了向注册中心注册的动作
                if (!register() ) {
                    throw new IllegalStateException("Registration error at startup. Invalid server response.");
                }
            } catch (Throwable th) {
                logger.error("Registration error at startup: {}", th.getMessage());
                throw new IllegalStateException(th);
            }
        }

        // 启动定时任务（如解析集群、开启心跳、复制和获取实例信息）
        initScheduledTasks();

        try {
            Monitors.registerObject(this);
        } catch (Throwable e) {
            logger.warn("Cannot register timers", e);
        }

        // 使用 DiscoveryManager.getInstance()获取实例
        // 使用 DiscoveryClient
        DiscoveryManager.getInstance().setDiscoveryClient(this);
        DiscoveryManager.getInstance().setEurekaClientConfig(config);

        initTimestampMs = System.currentTimeMillis();
        initRegistrySize = this.getApplications().size();
        registrySize = initRegistrySize;
        logger.info("Discovery Client initialized at timestamp {} with initial instances count: {}", initTimestampMs, initRegistrySize);
    }
```

下面的代码用来初始化心跳任务。

```
heartbeatExecutor = new ThreadPoolExecutor(
    1, clientConfig.getHeartbeatExecutorThreadPoolSize(), 0, TimeUnit.SECONDS,
    new SynchronousQueue<Runnable>(),
    new ThreadFactoryBuilder().setNameFormat("DiscoveryClient-HeartbeatExecutor-%d").setDaemon(true).build()
    );  // 使用直接传递
```

初始化本地缓存刷新任务。

```
cacheRefreshExecutor = new ThreadPoolExecutor(
    1, clientConfig.getCacheRefreshExecutorThreadPoolSize(), 0, TimeUnit.SECONDS,
    new SynchronousQueue<Runnable>(),
    new ThreadFactoryBuilder().setNameFormat("DiscoveryClient-CacheRefreshExecutor-%d").setDaemon(true).build()
    );
```

拉取注册表动作。

```
fetchRegistry(false)
```

启动定时任务。

```
initScheduledTasks();
```

执行从 Eureka Server 拉取注册表的任务。

```
boolean primaryFetchRegistryResult = fetchRegistry(false);
```

执行注册任务。

```
register();
```

2.4　Nacos 的使用

Nacos 是阿里巴巴公司开发的一个框架，用于在微服务中进行服务的发现、配置和管理。它不仅可以作为注册中心，还可以作为配置中心等。本节讲解它的注册中心的功能。

下面先搭建 Nacos 的 Server 端。

2.4.1 搭建单节点 Nacos Server

单节点 Nacos Server 的搭建步骤如下。

（1）下载 Nacos 组件。

打开 Nacos 的 github 地址 https://github.com/alibaba/nacos/releases。下载 nacos-server-1.1.4.tar.gz 并解压。

（2）启动 Nacos。

在 Windows 系统下，执行下面的启动命令（standalone 代表单机模式运行，非集群模式）。

```
cmd startup.cmd -m standalone
```

Nacos 启动界面如图 2-17 所示。

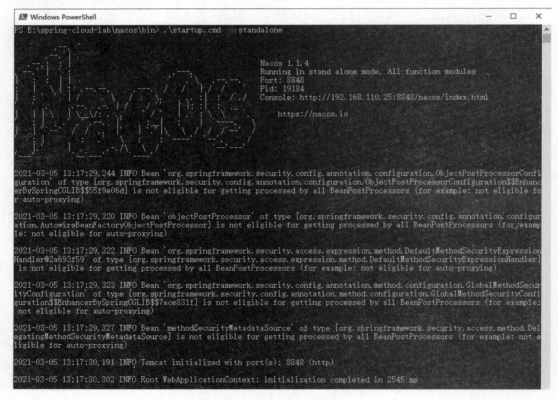

图 2-17　Nacos 启动界面

第 2 章 微服务注册中心

能看到如图 2-18 所示的结果，就表示 Nacos 启动成功。

```
2021-03-05 13:17:33,563 INFO Exposing 2 endpoint(s) beneath base path '/actuator'
2021-03-05 13:17:33,586 INFO Initializing ExecutorService 'taskScheduler'
2021-03-05 13:17:33,666 INFO Tomcat started on port(s): 8848 (http) with context path '/nacos'
2021-03-05 13:17:33,671 INFO Nacos Log files: E:\spring-cloud-lab\nacos\logs\
2021-03-05 13:17:33,671 INFO Nacos Conf files: E:\spring-cloud-lab\nacos\conf\
2021-03-05 13:17:33,672 INFO Nacos Data files: E:\spring-cloud-lab\nacos\data\
2021-03-05 13:17:33,672 INFO Nacos started successfully in stand alone mode.
2021-03-05 13:17:37,580 INFO Initializing Servlet 'dispatcherServlet'
2021-03-05 13:17:37,588 INFO Completed initialization in 8 ms
```

图 2-18　Nacos 启动成功界面

在 Linux/Unix/Mac 系统下，执行下面的启动命令（standalone 代表单机模式运行，非集群模式）。读者可以自己运行。

```
sh startup.sh -m standalone
```

（3）访问 Nacos。

在浏览器中访问 http://192.168.110.25:8848/nacos/index.html，在打开的页面中输入用户名 nacos 和密码 nacos 即可登录。

在页面左侧选项栏中选择"服务管理"→"服务列表"选项，在页面右侧将显示 Nacos 作为服务注册中心的管理界面，如图 2-19 所示。至此 Nacos 启动成功。

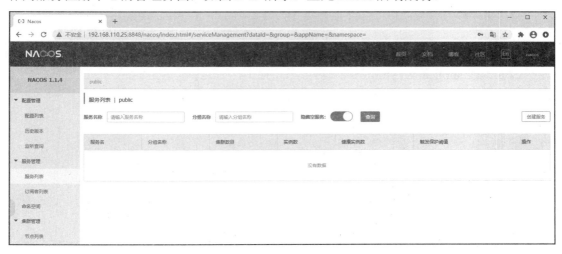

图 2-19　Nacos 服务注册中心管理界面

2.4.2 创建 Nacos Client

先创建一个 Spring Boot 项目 nacos-client。

(1) 在 pom 中添加如下依赖。

```xml
<dependencies>
    <dependency>
        <groupId>org.springframework.boot</groupId>
        <artifactId>spring-boot-starter-web</artifactId>
    </dependency>

    <dependency>
        <groupId>org.springframework.boot</groupId>
        <artifactId>spring-boot-starter-test</artifactId>
        <scope>test</scope>
    </dependency>

    <!-- nacos 服务发现依赖 -->
    <dependency>
        <groupId>com.alibaba.cloud</groupId>
        <artifactId>spring-cloud-starter-alibaba-nacos-discovery</artifactId>
        <version>2.2.3.RELEASE</version>
    </dependency>

</dependencies>
```

(2) 修改 application.yml。

```
server:
  port: 8080
spring:
  application:
    name: nacos-client
  cloud:
    nacos:
      discovery:
        # 服务注册中心地址
        server-addr: 127.0.0.1:8848
```

（3）在启动类上增加注解。

```
@EnableDiscoveryClient
@SpringBootApplication
@EnableDiscoveryClient
public class NacosClientApplication {
    public static void main(String[] args) {
        SpringApplication.run(NacosClientApplication.class, args);
    }
}
```

启动服务，Nacos 服务列表如图 2-20 所示。

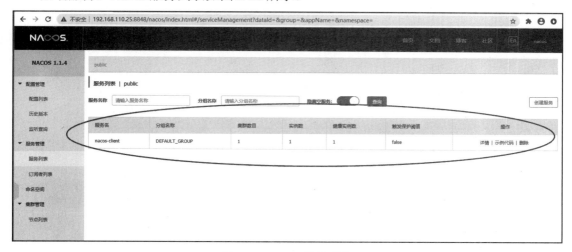

图 2-20　Nacos 服务列表

2.4.3　高可用 Nacos Server 搭建

将解压后的 nacos 复制 3 份，就用这 3 个节点搭建 Nacos Server 集群，如图 2-21 所示。

📁 nacos-cluster-8849	2021/3/5 15:08	文件夹
📁 nacos-cluster-8850	2021/3/5 15:08	文件夹
📁 nacos-cluster-8851	2021/3/5 15:08	文件夹

图 2-21　Nacos Server 集群

搭建高可用 Nacos Server 时，每个节点都需要修改以下 3 个地方。

(1) 修改 conf\application.properties 中的 server.port。注意：3 个节点如果存在一台机器上，需要修改为不同的端口，防止端口冲突。在这里将端口分别修改为 8849、8850、8851。

```
# 节点 nacos-cluster-8849 的端口为 8849
server.port=8849
```

(2) 修改每个节点中的 conf/cluster.conf 文件，配置 3 个节点的 IP 和端口。

```
192.168.110.25:8849
192.168.110.25:8850
192.168.110.25:8851
```

(3) 创建数据库，并修改配置。

在 MySQL 数据库中创建 nacos-test 库，并导入 conf 目录下的 nacos-mysql.sql 文件。执行操作之后，数据库中的表信息如图 2-22 所示。

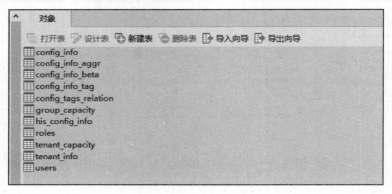

图 2-22　数据库中的表信息

修改每个节点中的 application.properties 文件，增加如下配置。

```
spring.datasource.platform=mysql
# 数据库实例数量
db.num=1
# 数据库连接信息
db.url.0=jdbc:mysql://localhost:3306/nacos_test?characterEncoding=utf8&connectTimeout=1000&socketTimeout=3000&autoReconnect=true
# 数据库用户名
db.user=root
# 数据库密码
```

```
db.password=root
```

启动每个节点,在每个节点执行如下命令。

```
startup.cmd -m cluster
```

访问每个节点的界面,如果能看到如图 2-23 所示信息,说明集群搭建成功。

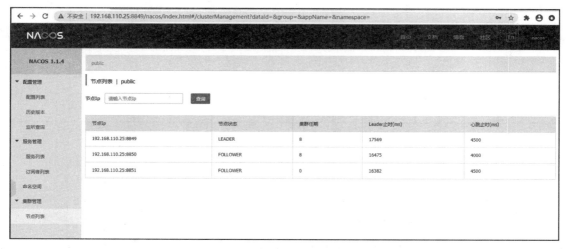

图 2-23　Nacos Server 集群搭建成功

从图 2-23 中可以看到 8849 节点为 LEADER,其他两个节点为 FOLLOWER。

集群搭建后,如果 Nacos Client 要接入集群也很简单,修改 application.yml 配置即可。

```
server:
  port: 8080
spring:
  application:
    name: nacos-client
  cloud:
    nacos:
      discovery:
        # 多个注册中心地址
        server-addr: 127.0.0.1:8849 ,127.0.0.1:8850,127.0.0.1:8851
```

启动 Nacos Client,分别查看 3 个节点的管理后台界面,然后观察 Nacos Server 的管理界面,会发现每个服务列表中都有一个 nacos-client 服务,如图 2-24 所示。

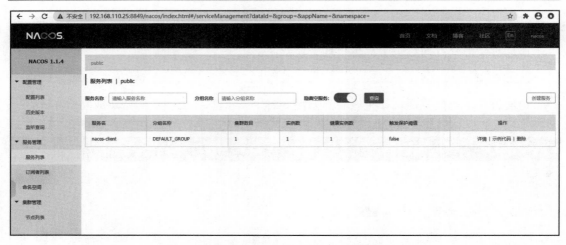

图 2-24　Nacos Server 管理界面中的服务列表

2.5　Consul 的使用

Consul 是 HashiCorp 公司推出的开源工具，它提供了服务发现、健康检查、Key/Value 存储等功能。本节只讨论它的服务注册发现功能。

2.5.1　搭建单节点 Consul Server

搭建单节点 Consul Server 的步骤如下。

（1）下载 Consul。

在浏览器中打开 https://www.consul.io/downloads，选择对应操作系统的 Consul 下载，笔者选择的是 Windows 系统，下载的文件是 consul_1.9.2_windows_amd64.zip。

（2）解压 Consul。

解压 consul_1.9.2_windows_amd64.zip 后得到一个 consul.exe 文件。

（3）启动。

在 Windows 系统上执行如下命令。

```
consul.exe agent -dev
```

在浏览器中访问 http://localhost:8500/ui/dc1/services，出现如图 2-25 所示界面，即表示 Consul 启动成功。

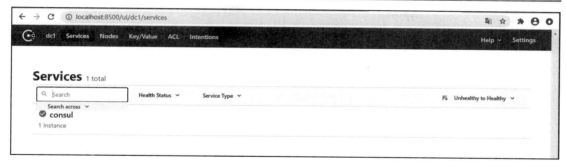

图 2-25　Consul 启动成功界面

2.5.2　创建 Consul Client

先创建一个 Spring Boot 项目 consul-client。

（1）在 pom 中添加如下依赖。

```
<dependencies>
    <dependency>
        <groupId>org.springframework.boot</groupId>
        <artifactId>spring-boot-starter-web</artifactId>
    </dependency>

    <dependency>
        <groupId>org.springframework.cloud</groupId>
        <artifactId>spring-cloud-starter-consul-discovery</artifactId>
    </dependency>

    <dependency>
        <groupId>org.springframework.boot</groupId>
        <artifactId>spring-boot-starter-test</artifactId>
        <scope>test</scope>
    </dependency>

    <dependency>
        <groupId>org.springframework.boot</groupId>
        <artifactId>spring-boot-starter-actuator</artifactId>
    </dependency>
</dependencies>
```

上面代码中添加了 spring-boot-starter-actuator 包，因为 Consul 需要做健康检查，如果没有这个包，在管理界面会出现一个红叉。

（2）修改 application.yml 文件。

```yaml
spring:
  cloud:
    consul:
      # 注册中心 IP
      host: localhost
      # 注册中心端口
      port: 8500
      discovery:
        # 是否注册到注册中心
        register: true
        # 实例 ID 唯一
        instance-id: ${spring.application.name}
        # 服务名称
        service-name: ${spring.application.name}
        # 端口
        port: ${server.port}


  application:
    name: consul-client

server:
  port: 8082
```

启动 consul-client 后看到如图 2-26 所示界面，即表示服务注册成功。

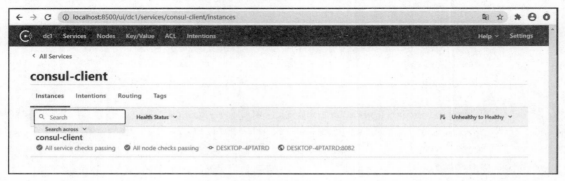

图 2-26　服务注册 Consul 成功界面

2.5.3 高可用 Consul Server 搭建

准备 3 台服务器,分别设为如下 IP 地址。

```
consul-01: 192.168.110.101
consul-02: 192.168.110.12
consul-03: 192.168.110.24
```

(1)在 3 台机器上安装 Consul。在每台机器上都执行如下命令,用于创建目录/usr/local/consul/。

```
mkdir -p /usr/local/consul/
```

下载 consul_1.9.2_linux_amd64.zip 文件到刚创建好的目录/usr/local/consul/。

```
wget https://releases.hashicorp.com/consul/1.9.2/consul_1.9.2_linux_amd64.zip
```

解压 consul_1.9.2_linux_amd64.zip,得到一个 Consul 可执行文件。

```
unzip consul_1.9.2_linux_amd64.zip -d /usr/local/consul/
```

创建 data 目录。

```
mkdir -p /usr/local/consul/data
```

以上操作在 3 台机器上都需要执行。

(2)启动 3 台机器上的 Consul。

执行以下命令,启动 consul-01 上的 Consul。

```
./consul agent -server -bind=192.168.110.101 -client=0.0.0.0 -ui -bootstrap-expect= 3 -data-dir=/usr/local/consul/data/ -node=server-01
```

该命令中参数的含义如下。

- ☑ -server:以服务端模式启动。
- ☑ -bind=192.168.110.101:当前 Consul 节点,绑定 IP192.168.110.101。
- ☑ -client=0.0.0.0:允许任何 IP 来向此节点注册。
- ☑ -ui:内在 Web 界面。
- ☑ -bootstrap-expect=3:期望有 3 个 Consul 节点。

- ☑ -data-dir=/usr/local/consul/data/：指定数据目录。
- ☑ -node=server-01：设置节点名称。

能看到如图 2-27 所示的界面，即表示启动成功。

图 2-27 Consul 节点 1 启动成功界面

执行以下命令，启动 consul-02 上的 Consul。

```
./consul agent -server -bind=192.168.110.103 -client=0.0.0.0 -ui -bootstrap-expect=3 -data-dir=/usr/local/consul/data/ -node=server-02
```

执行以下命令，启动 consul-03 上的 Consul。

```
./consul agent -server -bind=192.168.110.24 -client=0.0.0.0 -ui -bootstrap-expect=3 -data-dir=/usr/local/consul/data/ -node=server-03
```

（3）使 consul-02 和 consul-03 加入 consul-01（主节点）。

在 consul-02 和 consul-03 上，分别执行以下命令。

```
[root@localhost consul]# ./consul join 192.168.110.101
```

（4）访问 http://192.168.110.101:8500/ui/dc1/nodes。

能看到如图 2-28 所示界面，表示 Consul 集群启动成功。

集群启动成功后，如何让 consul client 将服务向它注册呢？consul client 并不会直接连接上面启动的 3 个节点。还需要在 consul client 所在的机器上执行如下命令，启动一个 consul agent。

第 2 章　微服务注册中心

```
D:\consul> .\consul.exe agent -client=0.0.0.0 -bind=192.168.110.25 -data-
dir=d:\consul\data -node=client-01
```

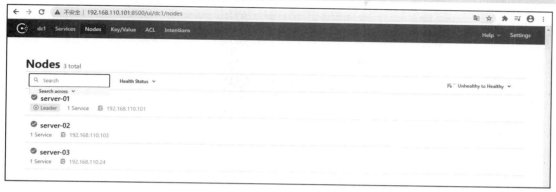

图 2-28　Consul 集群启动成功界面

该命令中参数的含义如下。

- ☑　agent：以 agent 模式启动。
- ☑　-client=0.0.0.0：允许任何 IP 来注册。
- ☑　-bind=192.168.110.25：绑定本机 IP，笔者的本机 IP 是 192.168.110.25。
- ☑　-data-dir=d:\consul\data：指定数据的目录。
- ☑　-node=client-01：设置节点名称。

启动 consul agent 后，再执行 join 命令，将此节点添加到 consul-01 上。

```
D:\consul\consul_1.9.2_windows_amd64>consul.exe join 192.168.110.101
```

在 Consul 的 3 个节点中的任意一个执行命令./consul members，可以看到这个命令是查找集群中的 Consul 节点信息，查询结果如图 2-29 所示。

```
[root@localhost consul]# ./consul members
Node       Address              Status  Type    Build  Protocol  DC   Segment
server-01  192.168.110.101:8301 alive   server  1.9.2  2         dc1  <all>
server-02  192.168.110.103:8301 alive   server  1.9.2  2         dc1  <all>
server-03  192.168.110.24:8301  alive   server  1.9.2  2         dc1  <all>
client-01  192.168.110.25:8301  alive   client  1.9.2  2         dc1  <default>
```

图 2-29　查询结果

至此，Consul 集群的 3 个 Server 节点和 consul agent 节点已经搭建完成。启动 2.5.2 节中的程序，即可向集群注册，注册结果如图 2-30 所示。

图 2-30 中的 consul-client 是业务代码，也就是正常的服务。在使用 Consul 时，consul

client 是不直接和 consul server 连接的。中间加了一层 consul agent，client 连接 agent，agent 再将连接分配到 server，注册逻辑如图 2-31 所示。

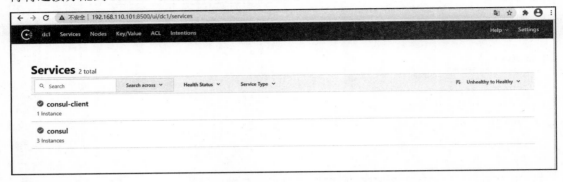

图 2-30　启动 consul client 后服务注册结果

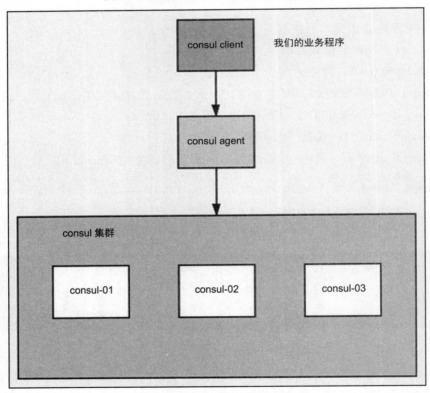

图 2-31　consul client 向集群注册逻辑图

2.6 小　　结

　　本章首先以生活中的通讯录类比注册中心，进行了注册中心思路的分析，通过分析总结出了注册中心的本质和原理，对注册中心的核心功能有了一定的认识。然后介绍了主流的注册中心 Eureka、Nacos、Consul。其中包括如何搭建单体注册中心、如何搭建集群注册中心，以及注册中心 client 如何向 server 注册等。由于 Eureka 是一款特别纯净的注册中心，对 Eureka Server 和 Eureka Client 的源码进行了学习，更深入地印证了前面分析的注册中心的本质和原理。其他注册中心 Nacos 和 Consul 的注册原理和核心功能与 Eureka 都是大同小异的。

第 3 章 服务调用

在一个微服务项目中，业务功能的实现是通过服务间彼此调用来实现的。通过第 2 章的学习，我们已经将服务注册到了注册中心。这章我们学习如何在服务之间完成彼此的调用。

3.1 生产环境中的微服务架构

在企业生产中，用户要请求后端微服务，后端的微服务一般分为两层，即业务层和能力层。其实业务层和能力层里的服务，就是普通的 Web 服务提供接口供其他服务调用。

业务层用于接收用户的请求，完成某个业务功能。而业务层如果要完成某个功能，则需要调用能力层的接口来实现。例如，有一项网约车业务，乘客支付订单（乘客支付业务流程图见图 3-1）。具体功能实现为：业务层的服务 api-passenger 接收乘客用户的支付请求，调用能力层的订单服务完成订单状态修改，从未支付修改为已支付，再调用能力层的支付服务完成订单的支付（这里的业务只是为了说明服务调用的关系，实际中的支付会借助第三方平台，如微信、支付宝）。

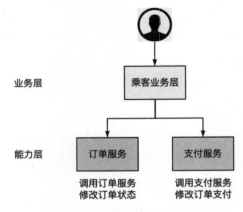

图 3-1 乘客支付业务流程图

Spring Cloud 提供了 RestTemplate 和 Feign 两种服务调用方式，下面分别进行介绍。

新建 passenger-api、order-service、pay-service 3 个项目，请参考第 2 章注册中心中的 eureka-client 的操作，这 3 个服务的端口分别是 8090、8091、8092。

Server 启动后，eureka 的管理界面如图 3-2 所示，可以看到 3 个服务已经注册成功。

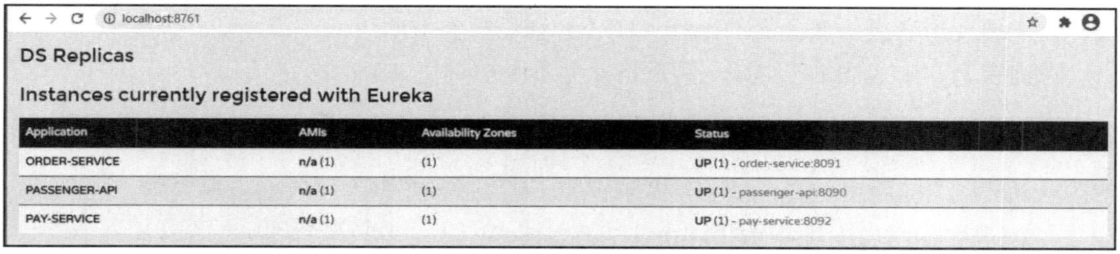

图 3-2　3 个服务注册成功

3.2　RestTemplate 调用

3.2.1　RESTful 风格介绍

曾经在面试工程师时，笔者发现很多求职者对 RESTful 风格有所误解，认为它是一个新的协议，其实它就是普通的 HTTP 请求。其中 REST 就是 resource representational state transfer 的简写，直接翻译即"资源表现层状态转移"。

resource 代表互联网资源。所谓"资源"就是网络上的一个实体，或者说是网上的一个具体信息，它可以是一段文本、一首歌曲、一种服务，可以使用一个 URI 指向它，每种"资源"对应一个 URI（uniform resource identifier，统一资源标识符）。

representational 是"表现层"。"资源"是一种消息实体，它可以有多种外在的表现形式，把"资源"具体呈现出来的形式叫作它的"表现层"。例如，文本可以用 TXT 格式表现，也可以使用 XML 格式、JSON 格式和二进制格式表现；视频可以用 MP4 格式表现，也可以用 AVI 格式表现。URI 只代表资源的实体，不代表它的形式。它的具体表现形式应该由 HTTP 请求的头信息 Accept 和 Content-Type 字段指定，这两个字段是对"表现层"的描述。

state transfer 是指"状态转移"。在客户端访问服务的过程中必然涉及数据和状态的转化，如果客户端要操作服务端资源，必须通过某种方法让服务器端资源发生"状态转移"。而这种转化是建立在表现层之上的，所以被称为"表现层状态转移"。客户端通过使用 HTTP

协议中常用的 4 个动词来实现上述操作，它们分别是获取资源的 GET、新建或更新资源的 POST、更新资源的 PUT 和删除资源的 DELETE。

RestTemplate 是由 Spring 提供的，用于封装 HTTP 调用，它可以简化客户端与 HTTP 服务器之间的交互，并且强制使用 RESTful 风格。RestTemplate 会处理 HTTP 的连接和关闭，只需要使用者提供服务器的地址（URL）和模板参数即可。

总而言之，RESTful 其实是一种风格，并不是一种协议。

大厂面试

面试官：能谈谈你对 RESTful 规范的理解吗？

面试者：resource 代表互联网资源，它是一个网络实体，可以是一段文本、一首歌曲、一种服务，每种资源对应一个 URI。representational 是"表现层"的意思。互联网资源有多种外在的表现形式，具体呈现出来的形式叫作"表现层"。例如，文本可以使用 TXT 格式，也可以使用 XML 格式、JSON 格式和二进制格式。URI 只代表资源的实体，不代表它的形式，它的具体表现形式由 HTTP 请求的头信息 Accept 和 Content-Type 字段指定，这两个字段是对"表现层"的描述。state transfer 是"状态转移"，客户端访问服务的过程中必然涉及数据和状态的转化。而这种转化是建立在表现层之上的，所以被称为"表现层状态转移"。客户端通过使用 HTTP 协议中常用的 4 个动作来实现上述操作，它们分别是获取资源的 GET、新建或更新资源的 POST、更新资源的 PUT 和删除资源的 DELETE。

3.2.2 RestTemplate 实战

其实 Spring 已经对 RESTful 风格的调用方式进行了封装，强烈推荐大家使用这种工具，具体的操作步骤如下。

（1）在 pay-service 中提供一个 restful 接口。

```
@RestController
@RequestMapping("/provider")
public class ProviderController {
    @GetMapping("/test")
    public String test() {
```

```
        return "pay restful provider";
    }
}
```

当其他服务来调用的时候，返回字符串 pay restful provider，如图 3-3 所示。

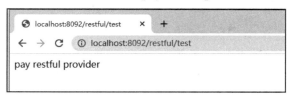

图 3-3　返回字符串

（2）在 passenger-api 中引入 RestTemplate，注意加一个注解@LoadBalanced。

```
@Bean
@LoadBalanced
public RestTemplate restTemplate(){
    return new RestTemplate();
}
```

（3）在 passenger-api 中写使用方法去调用 pay-service 中提供的接口。

```
@RestController
@RequestMapping("/restful")
public class RestTemplateController {
    @Autowired
    private RestTemplate restTemplate;
    @GetMapping("/test")
    public String test(){
        // 此处注意，一直写的 IP 和端口，现在是服务名（其实本质是虚拟主机名）:pay-service
        String url = "http://pay-service/provider/test";
        return restTemplate.getForObject(url, String.class);
    }
}
```

上面代码中@LoadBalanced 注解的作用，就是将 pay-service 解析成对应的 IP 和端口，这样就变成了一个普通的 HTTP 调用。

请求 http://localhost:8090/restful/test，返回结果如图 3-4 所示。

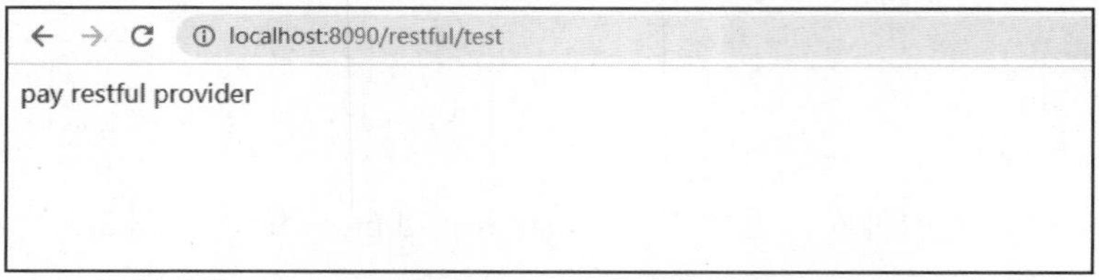

图 3-4　返回结果

通过图 3-4 可以看出 passenger-api 调用 pay-service 成功了。

3.2.3　RestTemplate 源码解析

下面以 restTemplate.getForObject 的实现源码为例,来解析 RestTemplate 的源码。打开源码可以看到如下代码。

```
@Nullable
public <T> T getForObject(String url, Class<T> responseType, Object...
uriVariables) throws RestClientException {
    RequestCallback requestCallback = this.acceptHeaderRequestCallback
(responseType);
    HttpMessageConverterExtractor<T> responseExtractor = new
HttpMessageConverterExtractor(responseType,this.getMessageConverters(),
this.logger);
    return this.execute(url, HttpMethod.GET, requestCallback,
responseExtractor, (Object[])uriVariables);
}
```

上述代码中设置了返回值类型,并且设置了 HttpMessageConverter。然后执行 execute 方法,在 execute 方法中设置 HTTP 请求的类型为 HttpMethod.GET。execute 方法的代码如下。

```
@Override
@Nullable
public <T> T execute(String url, HttpMethod method, @Nullable
RequestCallback requestCallback, @Nullable ResponseExtractor<T>
responseExtractor, Object... uriVariables) throws RestClientException {
    URI expanded = getUriTemplateHandler().expand(url, uriVariables);
    return doExecute(expanded, method, requestCallback, responseExtractor);
}
```

在 execute 方法中又执行了一个 doExecute 的方法。

```java
@Nullable
protected <T> T doExecute(URI url, @Nullable HttpMethod method, @Nullable
RequestCallback requestCallback, @Nullable ResponseExtractor<T>
responseExtractor) throws RestClientException {
    Assert.notNull(url, "URI is required");
    Assert.notNull(method, "HttpMethod is required");
    ClientHttpResponse response = null;

    Object var14;
    try {
        // 首先使用请求的 url 和 method (post 或者 get) 构造出一个 ClientHttpRequest
        ClientHttpRequest request = this.createRequest(url, method);
        if (requestCallback != null) {
            // 将之前的 requestBody requestHeader 放入此 ClientHttpRequest 中
            requestCallback.doWithRequest(request);
        }
        // 执行 execute 方法,获得 response
        response = request.execute();
        // 调用 handleResponse 方法处理 response 中存在的 error
        this.handleResponse(url, method, response);
        // 执行 extractData 方法将返回的 response 转换为某个特定的类型(这个类型是在
        调用之处设置的 response)
        var14 = responseExtractor != null ? responseExtractor.extractData
(response) : null;
    } catch (IOException var12) {
        String resource = url.toString();
        String query = url.getRawQuery();
        resource = query != null ? resource.substring(0, resource.
indexOf(63)) : resource;
        throw new ResourceAccessException("I/O error on " + method.name() +
" request for \"" + resource + "\": " + var12.getMessage(), var12);
    } finally {
        if (response != null) {
            // 关闭 ClientHttpResponse 资源
            response.close();
```

```
        }

    }

    return var14;
}
```

3.2.4 负载均衡

3.2.3 节的代码中用到了注解@LoadBalanced，这个注解的功能是将服务调用的 URL 中的服务名转换成 IP 地址。如果服务对应多个 IP，那么由负载均衡器将负载分担到多个 IP 上。

负载均衡器的原理是将注解@LoadBalanced 加到 RestTemplate 上，这样就会给 RestTemplate 访问的地址列表加一个拦截器，从而将请求拦截，然后按照一定的负载均衡规则，将请求路由到不同的服务上。

用 RestTemplate 执行请求时，有下面几个步骤。

（1）在参数中接收请求 URL（URL 中有服务名）。
（2）由框架来拦截请求，获取请求的 URL 中的服务名。
（3）通过服务名获取服务集合。
（4）通过负载均衡算法选择一个服务。
（5）将原来的 URL 中的服务名替换成上面选择后的服务的 IP 和端口。

3.2.5 自定义配置负载均衡

（1）在 pay-service 中定义一个接口，然后在 controller 中添加如下代码。

```
@Value("${server.port}")
private String port;

@GetMapping("/test-ribbon")
public String testRibbon(){
    System.out.println(port);
    return "pay restful provider";
}
```

（2）写 application-8092.yml 和 application-8093.yml 两个配置文件。

修改 application-8092.yml 文件，代码如下所示。

```yaml
spring:
  application:
    name: pay-service
server:
  port: 8092
eureka:
  client:
    serviceUrl:
      defaultZone: http://localhost:8761/eureka/
  instance:
    instance-id: ${spring.application.name}:${server.port}
```

修改 application-8093.yml 文件，代码如下所示。

```yaml
spring:
  application:
    name: pay-service
server:
  port: 8093
eureka:
  client:
    serviceUrl:
      defaultZone: http://localhost:8761/eureka/
  instance:
    instance-id: ${spring.application.name}:${server.port}
```

（3）启动两个 pay-service，端口分别是 8092 和 8093。启动后 Eureka 的管理界面如图 3-5 所示。

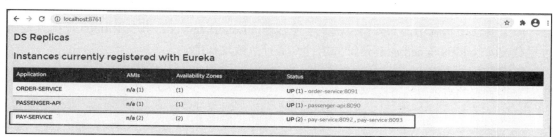

图 3-5　Eureka 的管理界面

（4）通过 passenger-api 来调用。调用 10 次后发现在 pay-server 的两个服务中，每个 pay-service 都接收到了 5 次请求。因为默认的负载均衡策略是 RoundRobinLoadBalancer，会轮循访问每个服务。

（5）配置负载均衡策略。在 passenger-api 中配置一个类 CustomLoadBalancerConfiguration。

```
public class CustomLoadBalancerConfiguration {

    @Bean
    ReactorLoadBalancer<ServiceInstance> randomLoadBalancer(Environment environment, LoadBalancerClientFactory loadBalancerClientFactory) {
        String name = environment.getProperty(LoadBalancerClientFactory.PROPERTY_NAME);
        return new RandomLoadBalancer(loadBalancerClientFactory
                .getLazyProvider(name, ServiceInstanceListSupplier.class),
            name);
    }
}
```

在配置类上增加一行配置。

```
@LoadBalancerClient(name = "pay-service",configuration = CustomLoadBalancerConfiguration.class)
public class MyConfiguration {
}
```

将负载策略改成随机负载策略后再调用 10 次，发现在 pay-server 的两个服务中每个 pay-service 接收到的请求次数不再是对半平分了，而是变成了随机接收。

3.2.6 Ribbon 源码解析

如果要使用负载均衡，就需要加入一个注解@LoadBalanced，其实 Ribbon 框架就是通过拦截被@LoadBalanced 注解修饰的组件来进行负载均衡的。

下面打开一个类 org.springframework.cloud.client.loadbalancer.LoadBalancerInterceptor，我们来查看它的方法。

```
@Override
public ClientHttpResponse intercept(final HttpRequest request, final byte[] body,
```

```
                final ClientHttpRequestExecution execution) throws IOException {
            final URI originalUri = request.getURI();
            String serviceName = originalUri.getHost();
            Assert.state(serviceName != null, "Request URI does not contain a valid hostname: " + originalUri);
                return this.loadBalancer.execute(serviceName, this.requestFactory.createRequest(request, body, execution));
        }
```

当调用远程服务时,可以执行下面的代码。

```
public String test(){
    // 此处注意,一直写的 IP 和端口,现在是服务名(其实本质是虚拟主机名):pay-service
    String url = "http://pay-service/provider/test";
    return restTemplate.getForObject(url, String.class);
}
```

当执行 getForObject 时会被 intercept 方法拦截下来。可以通过下面的代码解析出服务名(本质是虚拟主机名)。

```
String serviceName = originalUri.getHost();
```

下面进入 this.loadBalancer.execute 方法。

```
public <T> T execute(String serviceId, LoadBalancerRequest<T> request) throws IOException {
    String hint = this.getHint(serviceId);
    LoadBalancerRequestAdapter<T, DefaultRequestContext> lbRequest = new LoadBalancerRequestAdapter(request, new DefaultRequestContext(request, hint));
    Set<LoadBalancerLifecycle> supportedLifecycleProcessors = LoadBalancerLifecycleValidator.getSupportedLifecycleProcessors(this.loadBalancerClientFactory.getInstances(serviceId, LoadBalancerLifecycle.class), DefaultRequestContext.class, Object.class, ServiceInstance.class);
    supportedLifecycleProcessors.forEach((lifecycle) -> {
        lifecycle.onStart(lbRequest);
    });
    // 根据 serviceId 获取服务实例
```

```java
    ServiceInstance serviceInstance = this.choose(serviceId, lbRequest);
    if (serviceInstance == null) {
        supportedLifecycleProcessors.forEach((lifecycle) -> {
            lifecycle.onComplete(new CompletionContext(Status.DISCARD, lbRequest, new EmptyResponse()));
        });
        throw new IllegalStateException("No instances available for " + serviceId);
    } else {
        // 执行请求操作
        return this.execute(serviceId, serviceInstance, lbRequest);
    }
}
```

接着进入 execute 方法。

```java
@Override
public <T> T execute(String serviceId, ServiceInstance serviceInstance, LoadBalancerRequest<T> request) throws IOException {
    DefaultResponse defaultResponse = new DefaultResponse(serviceInstance);
    Set<LoadBalancerLifecycle> supportedLifecycleProcessors = LoadBalancerLifecycleValidator.getSupportedLifecycleProcessors
            (loadBalancerClientFactory.getInstances(serviceId, LoadBalancerLifecycle.class), DefaultRequestContext.class, Object.class, ServiceInstance.class);
    Request lbRequest = request instanceof Request ? (Request) request : new DefaultRequest<>();
    supportedLifecycleProcessors.forEach(lifecycle -> lifecycle.onStartRequest(lbRequest, new DefaultResponse(serviceInstance)));
    try {
        T response = request.apply(serviceInstance);
        Object clientResponse = getClientResponse(response);
        supportedLifecycleProcessors.forEach(lifecycle -> lifecycle.onComplete(new CompletionContext<>(CompletionContext.Status.SUCCESS, lbRequest, defaultResponse, clientResponse)));
        return response;
    }
    catch (IOException iOException) {
```

```
            supportedLifecycleProcessors.forEach(lifecycle -> lifecycle.
onComplete(new CompletionContext<>(CompletionContext.Status.FAILED,
iOException, lbRequest, defaultResponse)));
      throw iOException;
    }
    catch (Exception exception) {supportedLifecycleProcessors.forEach
(lifecycle -> lifecycle.onComplete(new CompletionContext<>(CompletionContext.
Status.FAILED, exception, lbRequest, defaultResponse)));
        ReflectionUtils.rethrowRuntimeException(exception);
    }
    return null;
}
```

然后进入如下方法。

```
@Override
public ClientHttpResponse execute(HttpRequest request, byte[] body) throws
IOException {
    if (this.iterator.hasNext()) {
        ClientHttpRequestInterceptor nextInterceptor = this.iterator.next();
        return nextInterceptor.intercept(request, body, this);
    }
    else {
        // 获取http方法
        HttpMethod method = request.getMethod();
        Assert.state(method != null, "No standard HTTP method");
        ClientHttpRequest delegate = requestFactory.createRequest(request.
getURI(), method);
        request.getHeaders().forEach((key, value) -> delegate.getHeaders().
addAll(key, value));
        if (body.length > 0) {
            if (delegate instanceof StreamingHttpOutputMessage) {
                StreamingHttpOutputMessage streamingOutputMessage =
(StreamingHttpOutputMessage) delegate;
                streamingOutputMessage.setBody(outputStream -> StreamUtils.
copy(body, outputStream));
            }
```

```
        else {
            StreamUtils.copy(body, delegate.getBody());
        }
    }
    return delegate.execute();
    }
}
```

通过 delegate.execute()方法一直跟踪，发现如下代码。

```
@Override
protected ClientHttpResponse executeInternal(HttpHeaders headers) throws IOException {
    HttpComponentsClientHttpRequest.addHeaders(this.httpRequest, headers);

    if (this.httpRequest instanceof HttpEntityEnclosingRequest && this.body != null) {
        HttpEntityEnclosingRequest entityEnclosingRequest = (HttpEntityEnclosingRequest) this.httpRequest;
        HttpEntity requestEntity = new StreamingHttpEntity(getHeaders(), this.body);
        entityEnclosingRequest.setEntity(requestEntity);
    }

    HttpResponse httpResponse = this.httpClient.execute(this.httpRequest, this.httpContext);
    return new HttpComponentsClientHttpResponse(httpResponse);
}
```

上述代码其实就是通过 HttpClient 完成了 HTTP 的调用。

总结，由于加了@LoadBalanced 注解，使用 RestTemplateCustomizer 对所有标注了注解@LoadBalanced 的 RestTemplate Bean 添加了一个 LoadBalancerInterceptor 拦截器。利用 RestTemplate 的拦截器，Spring 可以对 RestTemplate Bean 进行定制，加入 LoadBalancerInterceptor 拦截器进行从服务名到 ip:port 形式的替换，也就是将请求的地址中的服务逻辑名转为具体的服务地址，在进行服务名到服务地址的转换过程中，进行了负载均衡的操作。

> **大厂面试**
>
> 面试官：Ribbon 的原理是什么？为什么用一个注解@LoadBalanced 就能实现对服务的调用？
>
> 分析：重点回答以下 3 点。
>
> （1）通过拦截器对被注解@LoadBalanced 修饰的 RestTemplate 进行拦截。
>
> （2）将 RestTemplate 中调用的服务名解析成具体的 IP 地址，由于一个服务名会对应多个地址，因此在选择具体服务地址时，需要做负载均衡。
>
> （3）确定目标服务的 IP 和端口后，通过 HttpClient 进行 HTTP 的调用。

3.3 OpenFeign 调用

3.3.1 OpenFeign 的基础使用

Spring Cloud OpenFeign 是一个声明式的 HTTP 客户端，它简化了 HTTP 客户端的开发，使得编写 Web 服务的客户端变得更容易。OpenFeign 集成了 Eureka、CircuitBreaker、LoadBalancer。只需要创建一个接口并写一个注解，Spring Cloud OpenFeign 就能实现 HTTP 调用。

下面介绍如何在项目中使用 OpenFeign。

（1）在调用方的 pom 中添加依赖。

```xml
<dependency>
    <groupId>org.springframework.cloud</groupId>
    <artifactId>spring-cloud-starter-openfeign</artifactId>
</dependency>
```

（2）定义一个接口 PayServiceClient。每个@FeignClient 对应着一个服务，它是调用某个服务的一个抽象。

```
// 此处是要调用的服务名
@FeignClient("pay-service")
public interface PayServiceClient {
```

```java
// 此处是要调用服务的路径和调用的方法
@RequestMapping(method = RequestMethod.GET,value = "/provider/test")
String restfulTest();
}
```

(3) 在启动类上加注解 @EnableFeignClients。

```java
@EnableFeignClients
@SpringBootApplication
@EnableDiscoveryClient
// 开启feign调用
@EnableFeignClients
public class PassengerApiApplication {
    public static void main(String[] args) {
        SpringApplication.run(PassengerApiApplication.class, args);
    }
    @Bean
    @LoadBalanced
    public RestTemplate restTemplate(){
        return new RestTemplate();
    }
}
```

(4) 写一个测试接口。

```java
@RestController
@RequestMapping("/openfeign")
public class OpenFeignController {
    @Autowired
    private PayServiceClient payServiceClient;
    @GetMapping("/test")
    public String testFeign(){
        return payServiceClient.restfulTest();
    }
}
```

访问 http://localhost:8090/openfeign/test，输出结果如图 3-6 所示。

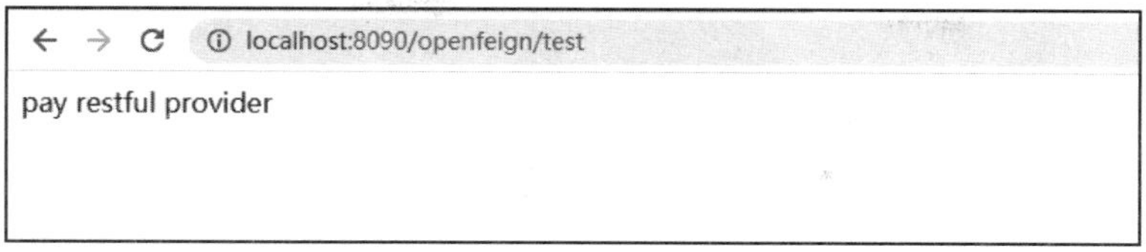

图 3-6 输出结果

3.3.2 自定义 URL

3.3.1 节中用到的@FeignClient("pay-service")，其中 pay-service 是系统的服务名（实际是虚拟主机名），也可以自定义成需要的名字，但是此时需要用一个属性 url 来指定它对应的实际地址。

```
@FeignClient(value = "my-pay-service",url = "http://localhost:8092/")
```

系统中并没有 my-pay-service 服务，但是通过 url 指定了它具体访问的路径，就可以对它进行访问了。

3.3.3 自定义 OpenFeign 的配置

下面以用户认证为例，通过配置实现如何在 OpenFeign 中增加用户名、密码，以实现服务的授权访问。

（1）改造 pay-service，让它的访问需要授权。在 pom 中引入 jar 包。

```xml
<dependency>
    <groupId>org.springframework.cloud</groupId>
    <artifactId>spring-cloud-starter-openfeign</artifactId>
</dependency>
```

修改 application.yml，配置用户名和密码。

```yaml
spring:
  security:
    user:
      name: root
      password: 123456
```

访问 http://localhost:8092/provider/test，查看返回结果发现需要登录。此时，用原来的 passenger-api 来请求 pay-service 时，会报 401 错误，因为没有添加授权信息，如图 3-7 所示。

```
feign.FeignException$Unauthorized: [401] during [GET] to [http://localhost:8092/provider/test] [PayServiceClient#restf
    at feign.FeignException.clientErrorStatus(FeignException.java:197) ~[feign-core-10.10.1.jar:na]
```

图 3-7 未授权错误

（2）修改 passenger-api。

上面报错的原因，是访问 pay-service 时需要用户名和密码，而 passenger-api 没有设置相应的用户名和密码。可以通过 OpenFeign 的自定义配置来设置。

新增一个配置类 AuthConfiguration。

```
public class AuthConfiguration {
    @Bean
    public BasicAuthRequestInterceptor basicAuthRequestInterceptor() {
        return new BasicAuthRequestInterceptor("root", "123456");
    }
}
```

在@FeignClient 中添加一个 configuration 属性。

```
@FeignClient(name = "my-pay-service",url = "http://localhost:8092/",
configuration = AuthConfiguration.class)
```

此时重启服务，再进行访问，即可访问成功。

3.3.4 Feign 源码解析

使用 Feign 的时候，在启动类上加了一个注解@EnableFeignClients，同时在每个调用的接口上写了注解@FeignClient。这样程序在启动后，就开启了对 Feign Client 的扫描。

打开 org.springframework.cloud.openfeign.FeignClientsRegistrar 类，可以看到 registerBeanDefinitions 方法。

```
public void registerBeanDefinitions(AnnotationMetadata metadata,
BeanDefinitionRegistry registry) {
    this.registerDefaultConfiguration(metadata, registry);
    this.registerFeignClients(metadata, registry);
}
```

再进入 registerDefaultConfiguration 方法，发现它的主要作用是获取@EnableFeignClients 注解的属性键-值对。

```
private void registerDefaultConfiguration(AnnotationMetadata metadata,
BeanDefinitionRegistry registry) {
    Map<String, Object> defaultAttrs = metadata.getAnnotationAttributes
(EnableFeignClients.class.getName(), true);
    if (defaultAttrs != null && defaultAttrs.containsKey
("defaultConfiguration")) {
        String name;
        if (metadata.hasEnclosingClass()) {
            name = "default." + metadata.getEnclosingClassName();
        } else {
            name = "default." + metadata.getClassName();
        }
        this.registerClientConfiguration(registry, name, defaultAttrs.get
("defaultConfiguration"));
    }
}
```

下面是 registerFeignClients 方法，其主要作用是对被注解@FeignClient 标记的接口进行处理。

```
public void registerFeignClients(AnnotationMetadata metadata,
BeanDefinitionRegistry registry) {
    LinkedHashSet<BeanDefinition> candidateComponents = new LinkedHashSet();
    Map<String, Object> attrs = metadata.getAnnotationAttributes
(EnableFeignClients.class.getName());
    Class<?>[] clients = attrs == null ? null : (Class[])((Class[])attrs.
get("clients"));
    if (clients != null && clients.length != 0) {
        Class[] var12 = clients;
        int var14 = clients.length;

        for(int var16 = 0; var16 < var14; ++var16) {
            Class<?> clazz = var12[var16];
            candidateComponents.add(new AnnotatedGenericBeanDefinition
(clazz));
        }
    } else {
        ClassPathScanningCandidateComponentProvider scanner = this.
getScanner();
```

```
        scanner.setResourceLoader(this.resourceLoader);
        scanner.addIncludeFilter(new AnnotationTypeFilter(FeignClient.class));
        Set<String> basePackages = this.getBasePackages(metadata);
        Iterator var8 = basePackages.iterator();
        while(var8.hasNext()) {
            String basePackage = (String)var8.next();
            candidateComponents.addAll(scanner.findCandidateComponents(basePackage));
        }
    }
    Iterator var13 = candidateComponents.iterator();
    while(var13.hasNext()) {
        BeanDefinition candidateComponent = (BeanDefinition)var13.next();
        if (candidateComponent instanceof AnnotatedBeanDefinition) {
            AnnotatedBeanDefinition beanDefinition = (AnnotatedBeanDefinition)candidateComponent;
            AnnotationMetadata annotationMetadata = beanDefinition.getMetadata();
            Assert.isTrue(annotationMetadata.isInterface(), "@FeignClient can only be specified on an interface");
            Map<String, Object> attributes = annotationMetadata.getAnnotationAttributes(FeignClient.class.getCanonicalName());
            String name = this.getClientName(attributes);
            this.registerClientConfiguration(registry, name, attributes.get("configuration"));
            this.registerFeignClient(registry, annotationMetadata, attributes);
        }
    }
}
```

上述代码调用了 registerFeignClient 方法，将 Bean 实例注册到 Spring 容器中，便于以后使用。

```
private void registerFeignClient(BeanDefinitionRegistry registry, AnnotationMetadata annotationMetadata, Map<String, Object> attributes) {
    String className = annotationMetadata.getClassName();
    Class clazz = ClassUtils.resolveClassName(className,(ClassLoader)null);
```

```java
    ConfigurableBeanFactory beanFactory = registry instanceof
ConfigurableBeanFactory ? (ConfigurableBeanFactory)registry : null;
    String contextId = this.getContextId(beanFactory, attributes);
    String name = this.getName(attributes);
    FeignClientFactoryBean factoryBean = new FeignClientFactoryBean();
    factoryBean.setBeanFactory(beanFactory);
    factoryBean.setName(name);
    factoryBean.setContextId(contextId);
    factoryBean.setType(clazz);
    BeanDefinitionBuilder definition = BeanDefinitionBuilder.
genericBeanDefinition(clazz, () -> {
        factoryBean.setUrl(this.getUrl(beanFactory, attributes));
        factoryBean.setPath(this.getPath(beanFactory, attributes));
        factoryBean.setDecode404(Boolean.parseBoolean(String.valueOf
(attributes.get("decode404"))));
        Object fallback = attributes.get("fallback");
        if (fallback != null) {
            factoryBean.setFallback(fallback instanceof Class ? (Class)
fallback : ClassUtils.resolveClassName(fallback.toString(),(ClassLoader)
null));
        }
        Object fallbackFactory = attributes.get("fallbackFactory");
        if (fallbackFactory != null) {
            factoryBean.setFallbackFactory(fallbackFactory instanceof
Class ? (Class)fallbackFactory : ClassUtils.resolveClassName
(fallbackFactory.toString(), (ClassLoader)null));
        }
        return factoryBean.getObject();
    });
    definition.setAutowireMode(2);
    definition.setLazyInit(true);
    this.validate(attributes);
    String alias = contextId + "FeignClient";
    AbstractBeanDefinition beanDefinition = definition.getBeanDefinition();
    beanDefinition.setAttribute("factoryBeanObjectType", className);
    beanDefinition.setAttribute("feignClientsRegistrarFactoryBean",
factoryBean);
    boolean primary = (Boolean)attributes.get("primary");
```

```
    beanDefinition.setPrimary(primary);
    String qualifier = this.getQualifier(attributes);
    if (StringUtils.hasText(qualifier)) {
        alias = qualifier;
    }
    BeanDefinitionHolder holder = new BeanDefinitionHolder(beanDefinition,
className, new String[]{alias});
    BeanDefinitionReaderUtils.registerBeanDefinition(holder, registry);
}
```

上述代码中，前面两个方法进行了 BeanDefinition 注册。下面来看在具体调用的过程中是如何将接口进行实例化的。因为接口无法进行实际的业务处理，所以需要对应的类实例来完成。

打开 org.springframework.cloud.openfeign.FeignClientFactoryBean 类，通过类中的 getObject() 方法，类可获取对应 Bean 的实例信息，此时的实例是指被 @FeignClient 修饰的接口类的实例。

```
public Object getObject() {
    return this.getTarget();
}
```

实际上，getObject()方法是通过 getTarget()方法来获取实例的。

```
<T> T getTarget() {
    FeignContext context = beanFactory != null ? beanFactory.getBean
(FeignContext.class): applicationContext.getBean(FeignContext.class);
    Feign.Builder builder = feign(context);
    if (!StringUtils.hasText(url)) {
        if (!name.startsWith("http")) {
            url = "http://" + name;
        }
        else {
            url = name;
        }
        url += cleanPath();
        return (T) loadBalance(builder, context, new HardCodedTarget<>
(type, name, url));
    }
    if (StringUtils.hasText(url) && !url.startsWith("http")) {
        url = "http://" + url;
```

```
    }
    String url = this.url + cleanPath();
    Client client = getOptional(context, Client.class);
    if (client != null) {
        if (client instanceof FeignBlockingLoadBalancerClient) {
            // 因为有 url，所以不是负载均衡
            // Spring Cloud LoadBalancer 位于类目录，所以没有做包装
            client = ((FeignBlockingLoadBalancerClient) client).getDelegate();
        }
        builder.client(client);
    }
    Targeter targeter = get(context, Targeter.class);
    return (T) targeter.target(this, builder, context, new HardCodedTarget<>
(type, name, url));
}
```

getTarget()方法中有一个 Feign.Builder，它的作用是负责生成被@FeignClient 修饰的接口类实例，通过 Java 的反射机制生成实例，当 feignclient 的方法被调用时，InvocationHandler 的回调函数会被调用。接下来看一下调用时的逻辑处理。打开 feign.SynchronousMethodHandler 类，有一个 invoke 方法。

```
@Override
public Object invoke(Object[] argv) throws Throwable {
    RequestTemplate template = buildTemplateFromArgs.create(argv);
    Options options = findOptions(argv);
    Retryer retryer = this.retryer.clone();
    while (true) {
        try {
            return executeAndDecode(template, options);
        } catch (RetryableException e) {
            try {
                retryer.continueOrPropagate(e);
            } catch (RetryableException th) {
                Throwable cause = th.getCause();
                if (propagationPolicy == UNWRAP && cause != null) {
                    throw cause;
                } else {
                    throw th;
```

```
            }
        }
        if (logLevel != Logger.Level.NONE) {
            logger.logRetry(metadata.configKey(), logLevel);
        }
        continue;
    }
}
```

利用 buildTemplateFromArgs.create(argv)构建了一个 RequestTemplate，打开 executeAndDecode 方法。

```
Object executeAndDecode(RequestTemplate template, Options options) throws Throwable {
    Request request = targetRequest(template);
    if (logLevel != Logger.Level.NONE) {
        logger.logRequest(metadata.configKey(), logLevel, request);
    }

    Response response;
    long start = System.nanoTime();
    try {
        response = client.execute(request, options);
        // 确保创建了请求
        response = response.toBuilder().request(request).requestTemplate(template).build();
    } catch (IOException e) {
    if (logLevel != Logger.Level.NONE) {
        logger.logIOException(metadata.configKey(), logLevel, e, elapsedTime(start));
    }
    throw errorExecuting(request, e);
    }
    long elapsedTime = TimeUnit.NANOSECONDS.toMillis(System.nanoTime() - start);

    if (decoder != null) return decoder.decode(response, metadata.
```

```
returnType());

   CompletableFuture<Object> resultFuture = new CompletableFuture<>();
   asyncResponseHandler.handleResponse(resultFuture, metadata.configKey(),
response, metadata.returnType(), elapsedTime);

   try {
      if (!resultFuture.isDone())
      throw new IllegalStateException("Response handling not done");
      return resultFuture.join();
   } catch (CompletionException e) {
      Throwable cause = e.getCause();
      if (cause != null)
      throw cause;
      throw e;
   }
}
```

在上述方法里执行了 response = client.execute(request, options)。下面以 FeignBlockingLoadBalancerClient 为例。

```
@Override
public Response execute(Request request, Request.Options options) throws
IOException {
   final URI originalUri = URI.create(request.url());
   String serviceId = originalUri.getHost();
   Assert.state(serviceId != null, "Request URI does not contain a
valid hostname: " + originalUri);
   String hint = getHint(serviceId);
   DefaultRequest<RequestDataContext> lbRequest = new DefaultRequest<>
(new RequestDataContext(buildRequestData(request), hint));
   Set<LoadBalancerLifecyßle> supportedLifecycleProcessors =
LoadBalancerLifecycleValidator
         .getSupportedLifecycleProcessors(
               oadBalancerClientFactory.getInstances(serviceId,
LoadBalancerLifecycle.class),
               RequestDataContext.class, ResponseData.class,
ServiceInstance.class);
   supportedLifecycleProcessors.forEach(lifecycle -> lifecycle.onStart
```

```java
(lbRequest));
    ServiceInstance instance = loadBalancerClient.choose(serviceId,
lbRequest);
    org.springframework.cloud.client.loadbalancer.Response
<ServiceInstance> lbResponse = new DefaultResponse(instance);
    if (instance == null) {
        String message = "Load balancer does not contain an instance for
the service " + serviceId;
        if (LOG.isWarnEnabled()) {
            LOG.warn(message);
        }
        supportedLifecycleProcessors.forEach(lifecycle -> lifecycle
            .onComplete(new CompletionContext<ResponseData, ServiceInstance,
RequestDataContext>(
                CompletionContext.Status.DISCARD, lbRequest, lbResponse)));
        return Response.builder().request(request).status(HttpStatus.
SERVICE_UNAVAILABLE.value())
            .body(message, StandardCharsets.UTF_8).build();
    }
    String reconstructedUrl = loadBalancerClient.reconstructURI(instance,
originalUri).toString();
    Request newRequest = buildRequest(request, reconstructedUrl);
    return executeWithLoadBalancerLifecycleProcessing(delegate, options,
newRequest, lbRequest, lbResponse, supportedLifecycleProcessors);
}
```

上述代码中有一行 String reconstructedUrl = loadBalancerClient.reconstructURI(instance, originalUri). toString();可实现跟踪到最深处，下面看一下实际的代码实现。

```java
private static URI doReconstructURI(ServiceInstance serviceInstance, URI
original) {
    String host = serviceInstance.getHost();
    String scheme = Optional.ofNullable(serviceInstance.getScheme()).
orElse(computeScheme(original, serviceInstance));
    int port = computePort(serviceInstance.getPort(), scheme);

    if (Objects.equals(host, original.getHost()) && port == original.
getPort() && Objects.equals(scheme, original.getScheme())) {
```

```
            return original;
    }

    boolean encoded = containsEncodedParts(original);
    return UriComponentsBuilder.fromUri(original).scheme(scheme).host
(host).port(port).build(encoded).toUri();
}
```

从上述代码中,可以看到 scheme 是协议(HTTP),host 是调用的目标主机地址,port 是端口号,这样就可以发起对目标服务的调用了。

另外,Feign 通过对两个关键注解@EnableFeignClients 和@FeignClient 的解析,最终完成了 HTTP 的调用。

大厂面试

面试官:能简单描述一下 Feign 的工作原理吗?为什么在一个接口上写一个注解@FeignClient 就实现了远程服务的调用?

面试者:关键有以下两点。

(1)在启动类上标记注解@EnableFeignClients,对被注解@FeignClient 修饰的接口进行扫描。

(2)动态地对被注解@FeignClient 修饰的接口进行实例化,然后通过实例化后的类,进行调用逻辑处理。

3.4 小　　结

本章从一个微服务的例子入手,讲解了服务之间调用的两种方式,即 RestTemplate 和 OpenFeign。RestTemplate 不用引入额外的 jar 包依赖,就可以直接发送 HTTP 调用请求,在 RestTemplate 调用方式中,还学习了负载均衡。OpenFeign 是 Spring 封装的声明式、模板化的 HTTP 客户端,可以更快捷、优雅地调用 HTTP API。除了学习两种调用方式,还对 Ribbon 和 OpenFeign 的源码进行了解析,让大家能更深入地理解技术底层的实现原理,便于更好地指导开发。

第 4 章 服务的熔断、降级和限流

服务之间既然存在调用，那就不可避免地会出现调用出错的情况。如果服务之间的调用出了问题，就应该采取相应的措施来解决问题。

- ☑ 熔断，这个词大家应该都不陌生。在生活中，当家里的电路发生故障出现短路时，会造成保险丝的熔断。在股市中，当股指的波动幅度达到某个点后，交易所为了控制风险，采取的暂停交易的措施，就是股市的熔断。总之一句话，熔断就是当系统发生故障时，系统实行的保护措施，有时候也称为过载保护。
- ☑ 降级，是系统发生熔断时采取的替换措施。当正常的服务没法正常使用时，稍微降低一下标准，将就使用，等服务正常了，再正常使用。
- ☑ 限流，当访问系统的流量超过系统的负载时，系统所采取的一种保护措施，以防止流量过大造成系统的崩溃。

本章来学习微服务中的熔断、降级、限流相关的知识。

4.1 熔断和降级的应用场景

下面是一个服务调用场景，C 服务和 D 服务调用 B 服务，B 服务调用 A 服务。在情况 1 中，服务正常调用。服务在运行过程中，A 服务发生故障（网络延时、服务异常、负载过大无法及时响应），系统变成了情况 2。由于 B 服务调用 A 服务，A 服务出故障，导致 B 服务调用 A 的代码处出现故障，此时 B 服务也出故障了，系统变成了情况 3。以此类推，系统最终发展成 A、B、C、D 所有的服务都出错了，导致整个系统崩塌。这就是雪崩效应，如图 4-1 所示。

微服务系统之间通过互相调用来实现业务功能，但每个系统都无法百分之百保证自身的运行不会出问题。在服务调用中，很可能面临依赖服务失效的问题（网络延时、服务异常、负载过大无法及时响应）而导致服务雪崩，这对于一个系统来说是灾难性的。因此需要一个能提供强大的容错能力的组件，当服务发生异常时能提供保护，把影响控制在较小的范围内，避免造成所有服务的雪崩。

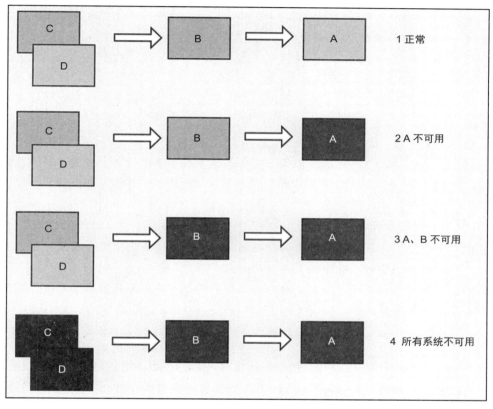

图 4-1　雪崩效应

熔断和降级需要一个参考标准，一般来说有以下两种。

（1）平均响应时间。例如，请求某个方法时，在某个时间段内（制定的指标），有 N（制定的指标）个请求的请求时间都超过了平均请求时间（制定的指标），那么在接下来的某个时间窗口内，对这个请求的调用将自动熔断。

（2）异常比例。例如，请求某个方法时，在某个时间段内（制定的指标），异常的数量超过了某个阈值（制定的指标），那么在接下来的某个时间窗口内，对这个请求的调用将自动熔断。

> ## 大厂面试
>
> 面试官：请描述一下熔断和降级的区别。
>
> 面试者：从两个方面来阐述。
>
> （1）熔断和降级的相似性。
> - ☑ 目的一致：都是从可用性和可靠性着想，为防止系统的整体响应缓慢甚至崩溃而采用的技术手段。
> - ☑ 最终表现类似：对于两者来说，最终让用户体验到的是某些功能暂时不可达或不可用。
> - ☑ 粒度一致：都是服务级别的。
> - ☑ 自治性要求很高：熔断模式一般都是服务基于策略的自动触发，降级虽说可人工干预，但在微服务架构下，完全靠人工显然不可能，开关预置、配置中心都是必要的手段。
>
> （2）熔断和降级的区别。
> - ☑ 触发原因不一样：服务的熔断一般是某个服务（下游服务）出现故障引起，而服务的降级一般是从整体负荷考虑。
> - ☑ 管理目标的层次不一样：熔断时每个服务都需要一个框架级的处理，而降级有业务层级之分，一般在服务调用的上层处理。

在 Spring Cloud 中最开始用的是 Hystrix 组件，但是 Hystrix 组件目前已经停止更新了，Spring Cloud 官方推荐使用 Resilience4j 组件提供熔断降级的功能，下面来学习该组件的使用。

4.2 熔断和降级的使用

熔断和降级主要是解决服务调用问题的，所以 4.1 节中分别从 RestTemplate 和 OpenFeign 两种服务调用方式来进行说明。本节还是使用上节的例子，利用服务 passenger-api 调用 pay-service。

4.2.1 RestTemplate 中熔断和降级的使用

（1）服务准备。pay-service 提供的接口，如图 4-2 所示。

图 4-2　pay-service 提供的接口

（2）修改 passenger-api，加入熔断组件 resilience4j。在 pom 中添加 resilience4j 的 jar 包。

```xml
<!-- 引入Resilience4J Circuit Breakers -->
<dependency>
    <groupId>org.springframework.cloud</groupId>
    <artifactId>spring-cloud-starter-circuitbreaker-resilience4j</artifactId>
</dependency>
```

修改调用 pay-service 接口的代码。

```java
@Autowired
private CircuitBreakerFactory circuitBreakerFactory;

@GetMapping("/test-resilience4j")
public String testResilience4j(){
    String url = "http://pay-service/provider/test";
    return circuitBreakerFactory.create("testResilience4j").run(()->
restTemplate.getForObject(url, String.class),throwable -> "熔断返回值");
}
```

上述代码 circuitBreakerFactory.create("testResilience4j")中的 testResilience4j 是熔断的 ID，可以根据这个 ID 获取熔断的配置（后面会讲，现在记住它是一个唯一标识）。

（3）下面测试熔断。

启动 pay-service 和 passenger-api。在浏览器中请求 http://localhost:8090/restful/test-resilience4j，返回结果如图 4-3 所示。

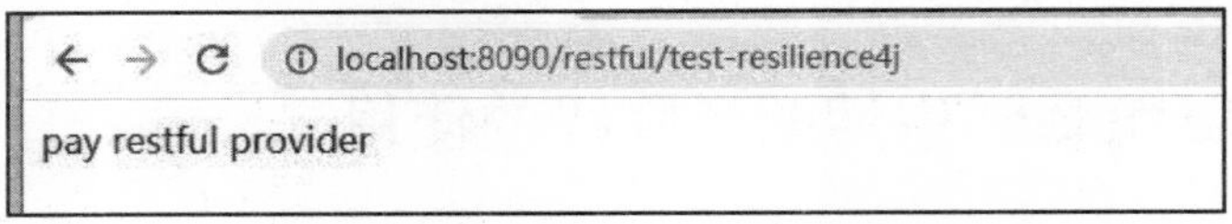

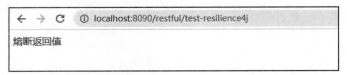

图 4-3　返回结果

停止 pay-service 服务，继续在浏览器中请求 http://localhost:8090/restful/test-resilience4j，返回结果如图 4-4 所示。

图 4-4　停止服务后的返回结果

4.2.2　OpenFeign 中熔断和降级的使用

在 OpenFeign 中熔断的使用和 RestTemplate 类似，代码如下。

```
@Autowired
private CircuitBreakerFactory circuitBreakerFactory;

@GetMapping("/test-resilience4j")
public String testResilience4j(){
    return circuitBreakerFactory.create("testFeign").run(()->
payServiceClient.restfulTest(),throwable -> "熔断返回值");
}
```

4.3　自定义熔断配置

可以通过在配置类中添加 Bean 来自定义熔断的配置。下面通过一个例子来讲解如何进行熔断配置。

在 pay-service 的接口中让程序睡眠 2s。

```
@GetMapping("/test")
public String test() {
    System.out.println(port);
```

```
    try {
        TimeUnit.SECONDS.sleep(2);
    } catch (InterruptedException e) {
        e.printStackTrace();
    }
    return "pay restful provider";
}
```

默认情况下，当 passenger-api 去请求 pay-service 时会触发熔断。在 passenger-api 中添加如下配置，将熔断超时时间修改成 4s。

```
@Bean
public Customizer<Resilience4JCircuitBreakerFactory> defaultCustomizer() {
    return factory -> factory.configureDefault(id -> new
Resilience4JConfigBuilder(id).timeLimiterConfig(TimeLimiterConfig.
custom().timeoutDuration(Duration.ofSeconds(4)).build()).
circuitBreakerConfig(CircuitBreakerConfig.ofDefaults()).build());
}
```

经测试发现程序不会触发熔断，返回了正常的结果。

上面配置的是通用的熔断策略，也可以定义个性化的熔断策略。在程序中添加一个 Bean。

```
@Bean
public Customizer<Resilience4JCircuitBreakerFactory> slowCustomizer() {
    return factory -> factory.configure(builder -> builder.
circuitBreakerConfig(CircuitBreakerConfig.ofDefaults()).
timeLimiterConfig(TimeLimiterConfig.custom().timeoutDuration
(Duration.ofSeconds(1)).build()), "slow");
}
```

代码中的"slow"就是此熔断策略的 ID，然后在应用的代码中做如下改动。

```
@GetMapping("/test-resilience4j")
public String testResilience4j(){
    String url = "http://pay-service/provider/test";
    return circuitBreakerFactory.create("slow").run(()->restTemplate.
getForObject(url, String.class),throwable -> "熔断返回值");
}
```

在 circuitBreakerFactory 的 create 方法中写上此 ID，个性化熔断配置就可以生效了。

4.4 限　　流

在互联网应用中，经常有突发导致流量过大的情况，如阿里系每年的"双 11"和"双 12"，京东商城每年的"618"，或者一些其他网站的秒杀活动。当这些场景发生时，最大的特点就是访问点瞬时会超出系统的处理能力。如果在没有任何保护机制的情况下，所有的流量都进入服务器，很可能造成服务器宕机，导致系统的不可用，继而带来经济损失。为了使系统在这些场景中稳定运行，除了前面学习的熔断和降级机制，还需要限流策略。

限流是通过限制并发访问数，或者限制一个时间窗口内允许处理的请求数量来保护系统，一旦达到限制的请求数量（略小于提前对系统进行压力测试得到的峰值），就对请求采取拒绝策略，如跳转到排队页面、提醒用户系统繁忙等。本质就是牺牲一部分用户的可用性，为大部分用户提供稳定的服务。

限流可以通过下面两种方式来实现。

（1）在 Nginx 层添加限流模块来进行限制。

（2）通过 Guava 提供的 Ratelimiter 来进行限制。

要实现限流，还需要对限流算法有所了解。这是面试中常考的问题，也是工作中常用的技术。

4.4.1 计数器（固定窗口）算法

计数器算法，是指在指定的时间周期内，累加访问的次数达到设定的阈值时，触发限流策略。在下一个时间周期进行访问时，访问次数清零。此算法无论在单机还是分布式环境下实现都非常简单，使用 Redis 的 incr 原子自增性，再结合 key 的过期时间，即可轻松实现，计算器算法如图 4-5 所示。

从图 4-5 看出，设置 1min 的阈值是 100，0:00～1:00 的请求数是 60，到 1:00 时请求数清零，再从 0 开始计算，这时 1:00～2:00 能处理的最大的请求数为 100，超过 100 个的请求，系统都拒绝。

这个算法有一个临界问题。例如，在图 4-5 中，0:00～1:00 只在 0:50 有 60 个请求，而 1:00～2:00 只在 1:10 有 60 个请求，虽然在两个 1min 的时间内都没有超过 100 个请求，但是 0:50～1:10 这 20s 内，却有 120 个请求，虽然在每个周期内都没超过阈值，但是在这 20s 内，已经远远超过了原来设置的 1min 内 100 个请求的阈值。

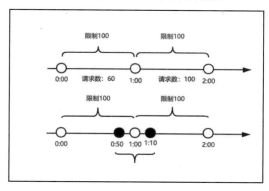

图 4-5　计数器算法

4.4.2　滑动时间窗口算法

为了解决计数器算法的临界值问题，开发了滑动窗口算法。在 TCP 网络通信协议中，就是采用滑动时间窗口算法来解决网络拥堵的问题。

滑动时间窗口是将计数器算法中的时间周期切分成多个小的时间窗口，分别在每个小的时间窗口中记录访问次数，然后根据时间将窗口往前滑动并删除过期的小时间窗口。最终只需要统计滑动窗口范围内的小时间窗口的总的请求数即可。滑动时间窗口算法如图 4-6 所示。

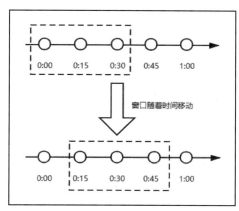

图 4-6　滑动时间窗口算法

在图 4-6 中，假设设置 1min 的请求阈值是 100，将 1min 拆分成 4 个小时间窗口，这样在每个小的时间窗口内，只能处理 25 个请求。用虚线方框表示滑动时间窗口，当前窗口的大小是 2，也就是在窗口内最多能处理 50 个请求。随着时间的推移，滑动窗口也随着时间往前移动。例如，图 4-6 中开始的时间时间窗口是 0:00～0:30，过了 15s 后，时间窗口是 0:15～0:45，时间窗口中的请求重新清零，这样就很好地解决了计数器算法的临界值问题。

在滑动时间窗口算法中，小窗口划分的越多，滑动窗口的滚动就越平滑，限流的统计就会越精确。

4.4.3 漏桶限流算法

漏桶限流算法的原理就像它的名字一样，维持一个漏斗，它有恒定的流出速度，不管水流流入的速度有多快，漏斗出水的速度始终保持不变，类似于消息中间件，不管消息的生产者请求量有多大，消息的处理能力取决于消费者。

漏桶的容量=漏桶的流出速度×可接受的等待时长。在这个容量范围内的请求可以排队等待系统的处理，超过这个容量的请求才会被抛弃。

在漏桶限流算法中，存在下面几种情况。

（1）当请求速度大于漏桶的流出速度时，也就是请求量大于当前服务所能处理的最大极限值时，触发限流策略。

（2）当请求速度小于或等于漏桶的流出速度时，也就是服务的处理能力大于或等于请求量时，正常执行。

漏桶算法有一个缺点，当系统在短时间内有突发的大流量时，漏桶算法就处理不了。

4.4.4 令牌桶限流算法

为了改进漏桶限流算法处理不了短期内的突发大流量的问题，引入了令牌桶限流算法。

令牌桶限流算法增加了一个大小固定的容器，也就是令牌桶，系统以恒定的速率向令牌桶中放入令牌，如果有客户端来请求，需要先从令牌桶中拿一个令牌，拿到了令牌才有资格访问系统，这时令牌桶中相应也少了一个令牌。当令牌桶满的时候，再向令牌桶生成令牌，令牌会被抛弃。

在令牌桶限流算法中，存在以下几种情况。

（1）当请求速度大于令牌的生成速度时，令牌桶中的令牌会被取完，后续再进来的请求，由于拿不到令牌，会被限流。

（2）当请求速度等于令牌的生成速度时，系统处于平稳状态。

（3）当请求速度小于令牌的生成速度时，系统的访问量远远低于系统的并发能力，请求可以被正常处理。

令牌桶限流算法由于有一个桶的存在，所以可以处理短时间内大流量的场景。这是令牌桶限流和漏桶限流算法的一个区别。

4.5 Sentinel 熔断和限流实战

Sentinel 是 Alibaba（后面简称阿里）开发的分布式微服务架构中的一款轻量级的熔断限流控制组件，它在阿里内部已经被广泛使用，经历过每年"双 11""双 12"的考验，现在是一款开源软件，也被很多互联网公司使用。

Sentinel 的官网地址为 https://github.com/alibaba/Sentinel。

使用 Sentinel 的目的，是通过一定的熔断限流规则来保护系统资源。资源可以是任何东西，服务、服务里的方法，甚至是一段代码。使用 Sentinel 来进行资源保护，主要分为以下几个步骤。

（1）定义资源。

（2）定义规则。

（3）检验规则是否生效。

先把可能需要保护的资源定义好（埋点），之后再配置规则。也可以这样理解，只要有了资源，就可以在任何时候灵活地定义各种流量控制规则。在编码的时候，只需要考虑该代码是否需要保护，如果需要保护，就将之定义为一个资源。

Sentinel 的使用可以分为以下两个部分。

- ☑ 核心库（Java 客户端）：不依赖任何框架/库，能够运行于 Java 7 及以上版本的运行时环境，同时对 Dubbo/Spring Cloud 等框架也有较好的支持。
- ☑ 控制台（Dashboard）：控制台主要负责管理推送规则、监控、集群限流分配管理、机器发现等。

4.5.1 Sentinel 控制台安装

从 https://github.com/alibaba/Sentinel/releases 下载 Sentinel 安装包。下载界面如图 4-7 所示。

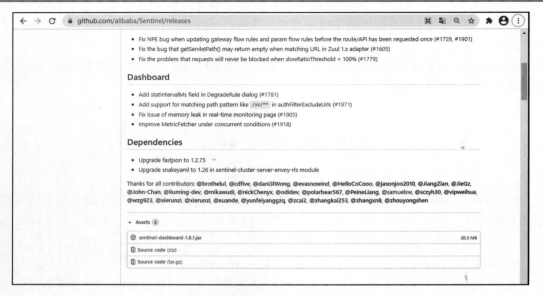

图 4-7 Sentinel 下载界面

下载的文件为 jar:sentinel-dashboard-1.8.1.jar。

执行如下命令。

```
java -jar -Dserver.port=18080 sentinel-dashboard-1.8.1.jar
```

-Dserver.port=18080：指定端口为：18080。

Sentinel 启动成功后，访问界面如图 4-8 所示。

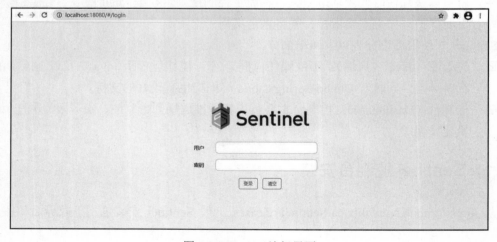

图 4-8 Sentinel 访问界面

输入用户名和密码(sentinel/sentinel),登录后界面如图 4-9 所示。

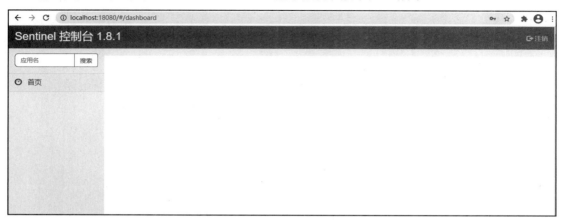

图 4-9 Sentinel 登录后界面

4.5.2 Sentinel 在程序中的配置

Sentinel 在程序中的配置步骤如下。

(1)新建一个项目 sentinel-local。

(2)修改 pom 文件。

```xml
<?xml version="1.0" encoding="UTF-8"?>
<project xmlns="http://maven.apache.org/POM/4.0.0" xmlns:xsi="http://
www.w3.org/2001/XMLSchema-instance" xsi:schemaLocation="http://maven.
apache.org/POM/4.0.0 https://maven.apache.org/xsd/maven-4.0.0.xsd">
    <modelVersion>4.0.0</modelVersion>
    <parent>
        <groupId>org.springframework.boot</groupId>
        <artifactId>spring-boot-starter-parent</artifactId>
        <version>2.4.3</version>
        <relativePath/> <!-- lookup parent from repository -->
    </parent>
    <groupId>com.cpf</groupId>
    <artifactId>sentinel-local</artifactId>
    <version>0.0.1-SNAPSHOT</version>
    <name>sentinel-local</name>
```

```xml
<description>Demo project for Spring Boot</description>

<properties>
    <java.version>1.8</java.version>
    <spring-cloud.version>2020.0.1</spring-cloud.version>
</properties>

<dependencies>
    <dependency>
        <groupId>org.springframework.boot</groupId>
        <artifactId>spring-boot-starter-web</artifactId>
    </dependency>

    <dependency>
        <groupId>org.springframework.boot</groupId>
        <artifactId>spring-boot-starter-test</artifactId>
        <scope>test</scope>
    </dependency>

<!-- https://mvnrepository.com/artifact/com.alibaba.cloud/spring-cloud-starter-alibaba-sentinel -->
    <dependency>
        <groupId>com.alibaba.cloud</groupId>
        <artifactId>spring-cloud-starter-alibaba-sentinel</artifactId>
        <version>2021.1</version>
    </dependency>

</dependencies>

<dependencyManagement>
    <dependencies>
        <dependency>
            <groupId>org.springframework.cloud</groupId>
            <artifactId>spring-cloud-dependencies</artifactId>
            <version>${spring-cloud.version}</version>
            <type>pom</type>
```

```xml
            <scope>import</scope>
        </dependency>
    </dependencies>
</dependencyManagement>

<build>
    <plugins>
        <plugin>
            <groupId>org.springframework.boot</groupId>
            <artifactId>spring-boot-maven-plugin</artifactId>
        </plugin>
    </plugins>
</build>
</project>
```

此处重要的是 com.alibaba.cloud:spring-cloud-starter-alibaba-sentinel:2021.1，通过这个 jar 包可以在项目中引入 Sentinel 组件。

（3）编写 application.yml。

```yml
server:
  port: 8091
spring:
  application:
    name: sentinel-local
  cloud:
    sentinel:
      eager: true
      transport:
        dashboard: localhost:18080
```

（4）编写代码，首先定义一个 Service，代码如下。

```
package com.cpf.sentinel.local.service;

import com.alibaba.csp.sentinel.annotation.SentinelResource;
import org.springframework.stereotype.Service;
```

```java
/**
 * @author 马士兵教育:chaopengfei
 * @date 2022/3/19
 */
@Service
public class SentinelLocalService {

    public String sayHello(){
        System.out.println("hello sentinel");
        return "hello sentinel";
    }

    public String sayHelloFail(){
        System.out.println("hello sentinel fail");
        return "hello sentinel fail";
    }
}
```

再定义一个 Controller，代码如下。

```java
package com.cpf.sentinel.local.controller;

import com.cpf.sentinel.local.service.SentinelLocalService;
import org.springframework.beans.factory.annotation.Autowired;
import org.springframework.web.bind.annotation.GetMapping;
import org.springframework.web.bind.annotation.RequestMapping;
import org.springframework.web.bind.annotation.RestController;

/**
 * @author 马士兵教育:chaopengfei
 * @date 2022/3/19
 */
@RestController
@RequestMapping("/sentinel")
public class SentinelController {

    @Autowired
```

```
SentinelLocalService sentinelLocalService;

@GetMapping("/test-local")
public String testLocal(){

    return sentinelLocalService.sayHello();

}
}
```

（5）请求接口 http://localhost:8091/sentinel/test-local。观察到服务出现在 Sentinel 界面中，如图 4-10 所示。

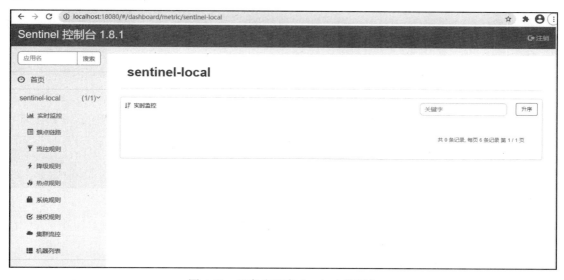

图 4-10　服务出现在 Sentinel 界面中

4.5.3　Sentinel 流控规则

1. 直接模式

首先单击 Sentinel 控制台左侧的"簇点链路"，然后单击右边的列表视图，接着单击每个接口右侧的"流控"按钮。设置 QPS（每秒查询数）的"单机阈值"为 2，如图 4-11 所示。

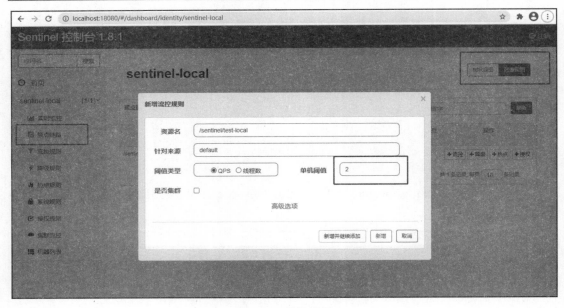

图 4-11 流控设置

下面通过 JMeter 工具来进行压力测试。设置 1s 发送 10 个请求数，如图 4-12 所示。

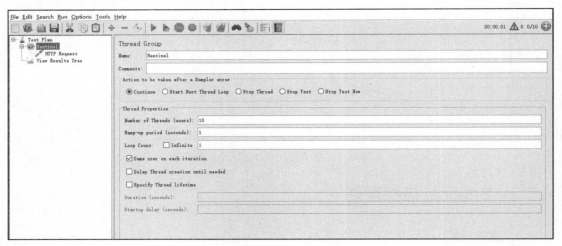

图 4-12 JMeter 设置

发送请求，查看结果，如图 4-13 所示。

从图 4-13 中可以发现，差不多 1s 只有两个请求成功。

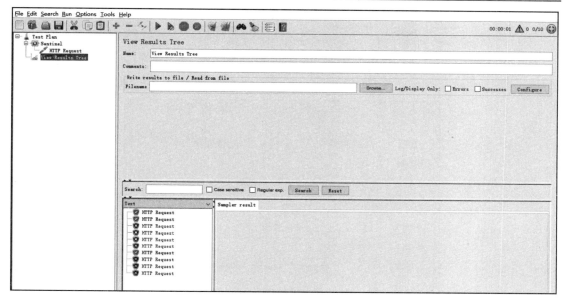

图 4-13 JMeter 流控结果

上面设置的是 QPS 限制,大家在设置流控的界面上可以看到,还有一个线程数的限制。QPS 限制和线程数的限制区别如下。

- ☑ QPS:请求在进入服务之前限制。
- ☑ 线程数:请求到达服务后,当服务线程处理不完时进行限制。

线程数的限制可以用程序睡眠来模拟。

```
@GetMapping("/test-local")
public String testLocal(){
    try {
        TimeUnit.SECONDS.sleep(1);
    } catch (InterruptedException e) {
        e.printStackTrace();
    }
    eturn sentinelLocalService.sayHello();
}
```

2. 关联模式

QPS 限流默认是直接模式,此模式只在当前访问的接口中生效。下面来看一下关联模式，关联是指一个接口是否触发限流规则，取决于另外一个接口。例如，在生产环境中，当扣减库存的服务达到阈值时，会触发下单服务的限流机制。

此时设置的配置，就是当关联的接口达到配置的阈值时，对主接口进行限流。如图 4-14 所示，当 /sentinel/test-local 接口达到 QPS 为 1 时，限制 /sentinel/test-local2 的访问。

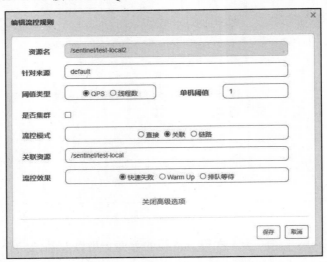

图 4-14　Sentinel 关联限流

3．流控效果

流控效果有以下 3 种，这 3 种效果只在阈值类型为 QPS 时才会出现。

- ☑ 快速失败：当触发流控规则时，直接报异常。
- ☑ Warm Up：会有一个预热时长选项，它指的是从最低限流次数（流次数/冷加载因子）开始，经过预热时长，才达到设置的 QPS 的值。
- ☑ 排队等待：匀速排队，让请求匀速通过，排队等待需让设置一个超时时间，超过这个时间，则触发流控规则。

4.5.4　Sentinel 降级规则

降级是指正常服务不可用时，在一个时间周期内，不再去调用正常的服务。Sentinel 的降级规则中有以下 3 种策略。

- ☑ 慢调用比例：需要设置参数，包括最大 RT（指响应时间，超过这个值则为慢应用）、比例阈值（慢调用占所有调用的比例）、熔断时长（在这段时间内发生熔断，拒绝

第 4 章 服务的熔断、降级和限流

所有的请求）、最小请求数（即允许通过的最小请求数，在该数量内不发生熔断）、统计时长（统计的时长周期）。

☑ 异常比例：需要设置参数，包括异常比例阈值（发生异常的请求数占总请求数的比例）、熔断时长（在这段时间内发生熔断，拒绝所有请求）、最小请求数（即允许通过的最小请求数，在该数量内不发生熔断）、统计时长（统计的时长周期）。

☑ 异常数：需要设置参数，包括异常数（发生异常的请求数量）、熔断时长（在这段时间内发生熔断，拒绝所有请求）、最小请求数（即允许通过的最小请求数，在该数量内不发生熔断）、统计时长（统计的时长周期）。

4.5.5 Sentinel 热点规则

还可以针对具体的请求进行流控规则的设置，即针对热点请求进行设置，热点规则如图 4-15 所示。

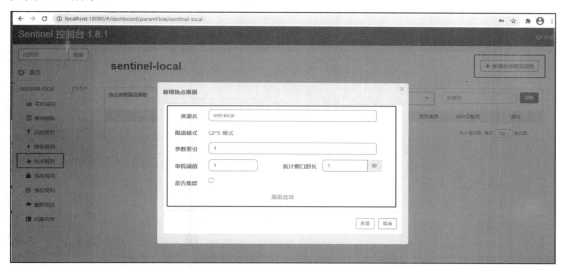

图 4-15 热点规则

在程序中，编写如下代码。

```
@GetMapping("/test-local")
@SentinelResource(value = "test-local",blockHandler = "testLocalFail")
public String testLocal(@RequestParam String p1){
    return sentinelLocalService.sayHello();
```

```
}
public String testLocalFail(String p1, BlockException e){
    return "异常了";
}
```

在上面代码中，加了一个注解@SentinelResource(value = "test-local",blockHandler = "testLocalFail")，资源名就是其中定义的 value 值，当触发热点规则时，请求参数中带参数 1（p1），如果触发 QPS 大于 1，就执行 testLocalFail()方法。

4.5.6 自定义流控处理

前面的内容中异常信息基本都是和正常的代码耦合在一起的，本节学习对代码无侵入的流控处理。

定义一个类 CustomHandler，代码如下。

```
import com.alibaba.csp.sentinel.slots.block.BlockException;

/**
 * 自定义流控处理类
 * @author 马士兵教育:chaopengfei
 * @date 2022/3/24
 */
public class CustomHandler {

    public static String handleException1(BlockException e1){
        return "异常1";
    }

    public static String handleException2(BlockException e1){
        return "异常2";
    }
}
```

在需要应用流控处理的地方配置一个注解。

```
@GetMapping("/test-custom")
@SentinelResource(value = "test-custom",blockHandlerClass = CustomHandler.class,blockHandler = "handleException1")
```

```
public String testLocal(){
    return "正常";
}
```

这样当服务触发限流时,就会进入类 CustomHandler 中的 handleException1 方法。

blockHandler 属性用于处理配置违规,只在触发流控规则时才会生效。fallback 属性用于处理 Java 异常。如果这两个属性同时存在,并且触发的条件都同时满足,则先运行 blockHandler 的处理逻辑。

4.6 小 结

本章主要讲解了在服务调用发生故障时,微服务提供的熔断、降级、限流的保障机制。以 RestTemplate 和 OpenFeign 两种服务调用方式为例,通过 resilience4j 组件进行了熔断降级的演示,也讲解了自定义熔断机制的配置方式。同时讲解了常用的限流算法、限流在 Sentinel 中的实战应用,以便于读者在生产实践中应用。

第 5 章

配 置 中 心

5.1 配置中心应用场景

在开发软件的过程中,有一些参数是需要配置在程序的代码外面(配置文件)的,如数据库连接池、功能性开关、注册中心地址等。随着业务的发展和程序的不断迭代,难免需要对配置进行添加、更新和删除等操作。大家试想一下,如果微服务特别多,有成千上万个,而且配置文件又散落在各个服务中,则是非常不方便的。这个时候就需要对配置进行统一管理,所以配置中心应运而生。

5.2 配置中心的设计思路

配置中心是必不可少的,本书也会在项目中使用到它。也有很多公司开源了非常优秀的配置中心可供使用,但是作为一名软件从业者,未来的技术精英,需要了解它的设计思路,这样在使用的时候,才能够更加得心应手。

我们换位思考,假设自己是配置中心的设计者,由自己来设计一个配置中心,那么需要从哪些方面来考虑呢?下面来一一介绍。

5.2.1 配置存储

平时使用的配置都是以 key-value 的形式来存储的。例如,在原来的代码中有下面的配置。

```
server:
  port: 8090
```

```yaml
spring:
  application:
    name: passenger-api
eureka:
  client:
    serviceUrl:
      defaultZone: http://localhost:8761/eureka/
  instance:
    instance-id: ${spring.application.name}:${server.port}
```

虽然用的是 yaml 文件格式，本质上其实还是 key-value 值。上面的配置相当于下面的配置，其本质是一样的。

```
server.port=8090
spring.application.name=passenger-api
eureka.client.serviceUrl.defaultZone=http://localhost:8761/eureka/
eureka.instance.instance-id=${spring.application.name}:${server.port}
```

那么这些配置应该存放到哪里呢？只要能存储数据的地方都可以，常用的有 Git、MySQL、File、内存、缓存等。

5.2.2 配置的属性

1．项目属性

我们知道了配置的基本存储是 key-value，那么在实际工作中关于配置的属性还有哪些呢？大家不要忘了，配置中心需要管理公司内部多个项目的配置，所以配置中心的配置要能区分具体的项目，也就是这个配置属于哪个项目。可以通过文件名来区分，也可以用目录来管理，有的配置中心还会通过对配置分组来实现等。

- ☑ 文件名管理的方式：如 projectA.yml 标识这个配置是项目 A 的，让文件的名称和项目名称一一对应。
- ☑ 目录管理的方式：可以通过创建一个项目目录来管理，此目录下都是该项目的配置文件。
- ☑ 分组管理的方式：是给配置文件增加一个分组属性，属于这个分组的配置文件都是这个项目的配置文件。

2. 环境属性

在实际工作中有很多环境，如开发环境、测试环境，还有预生产环境和生产环境。不同环境中的配置是不一样的，如数据库连接的地址就是不一样的，生产环境的数据库和本地开发的数据库肯定也不是一个库。

基于这个需求，就需要对配置所属的环境进行区分。可以参考项目属性的思路，通过文件名、目录和分组的方式来进行区分。

5.2.3 配置服务

前面分析了配置信息的存储结构、存储位置和配置的属性。现在我们基本就清楚了配置文件如何存储，以及在哪里存储了。那么这些存储好的配置如何提供给各个微服务使用呢？需要有一个服务去存储、更新配置，然后通过提供的 API 接口让配置使用方来查询。

还有一个更重要的步骤，就是配置变更后如何通知配置使用方。常用的方法是消息队列，当配置变更时向消息队列发送消息，配置使用方通过订阅消息来完成配置的动态刷新。

通过前面的分析，我们可以总结出配置中心需要下面几个功能。

- ☑ 按照正确的格式存储配置：如 key-value（键-值对）。
- ☑ 存储的位置：如 Git、MySQL、File、内存、缓存等。
- ☑ 配置中心服务端：从配置存储位置拉取配置，向配置使用方提供配置的查询接口。
- ☑ 配置使用方：从配置中心服务端查询配置，并使用配置。

配置中心的流程如图 5-1 所示。配置中心服务从存储读取配置，配置中心客户端（需要配置的业务服务）通过配置中心服务端来获取配置。

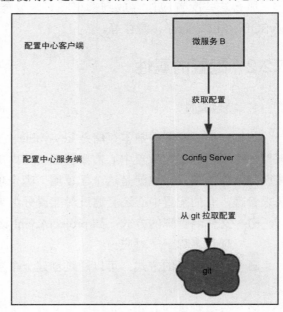

图 5-1 配置中心的流程

5.3 Spring Cloud 配置中心的使用

通过 Spring Cloud 的 config（配置中心）构建一个分布式配置中心特别简单。下面我们一步一步来学习。

5.3.1 在 Git 上创建配置

选择码云 Git 仓库，新建一个仓库 config-server，在仓库中创建一个文件 config-client-dev.yml，文件的内容如下。

```
testKey: testValue
```

5.3.2 创建配置的服务端

1. 准备工作

根据配置中心设计思路的分析，需要一个服务端来读取配置，并且提供配置查询的接口。创建一个 Spring Boot 的 Web 项目 cloud-config-server-git，并且将项目注册到 Eureka 注册中心，注册成功的界面如图 5-2 所示。

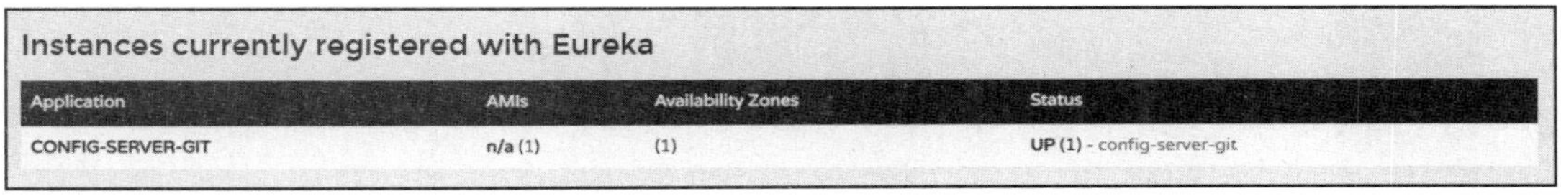

Instances currently registered with Eureka			
Application	AMIs	Availability Zones	Status
CONFIG-SERVER-GIT	n/a (1)	(1)	UP (1) - config-server-git

图 5-2　cloud-config-server-git 注册成功

2. 配置配置中心服务端

在上一步创建的 cloud-config-server-git 项目的 pom 文件中添加如下依赖。

```xml
<dependencies>
    <!-- 配置中心服务端：config-server -->
    <dependency>
        <groupId>org.springframework.cloud</groupId>
        <artifactId>spring-cloud-config-server</artifactId>
```

```xml
    </dependency>
    <dependency>
        <groupId>org.springframework.cloud</groupId>
        <artifactId>spring-cloud-starter-netflix-eureka-client</artifactId>
    </dependency>

    <dependency>
        <groupId>org.springframework.boot</groupId>
        <artifactId>spring-boot-starter-test</artifactId>
        <scope>test</scope>
    </dependency>
</dependencies>
```

注意：上述代码中多了一个 spring-cloud-config-server 的依赖。

修改 application.yml，增加配置中心的配置。

```yml
#服务端口
server:
  port: 6001

#应用名称及验证账号
spring:
  application:
    name: config-server-git
#配置中心配置
  cloud:
    config:
      server:
        git:
          uri: https://gitee仓库地址.git
          username: gitee账号
          password: gitee密码
          #访问超时时间
          timeout: 15

#注册中心
eureka:
```

```
  client:
    #设置服务注册中心的URL
    service-url:
      defaultZone: http://peer1:8761/eureka/
  instance:
    hostname: localhost
    instance-id: config-server-git
```

在启动类上增加一个注解@EnableConfigServer。

```
@EnableConfigServer
@SpringBootApplication
public class ConfigServerApplication {

    public static void main(String[] args) {
        SpringApplication.run(ConfigServerApplication.class, args);
    }

}
```

重启服务。访问 http://localhost:6001/config-client-dev.yml，如图 5-3 所示，配置中心获取 gitee 上的配置成功。

图 5-3　配置中心获取 gitee 上的配置成功

访问上面的地址和访问 http://localhost:6001/master/config-client-dev.yml 的结果是一样的。因为配置在 Git 上，所以和 Git 仓库对应，要有一个分支的概念，默认是 master 分支，这个在配置中心称作 label。访问地址后面的 dev，是用来区分环境的，如 dev 是开发环境、qa 是测试环境、pre 是预生产环境、prd 是生成环境等。下面新增一个 qa 测试环境来测试一下。

在 gitee 上新建一个文件 config-client-qa.yml，内容如下。

```
testKey: test-env
```

访问 http://localhost:6001/config-client-qa.yml，得到的测试环境结果如图 5-4 所示。

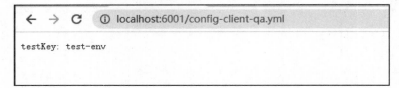

图 5-4　测试环境结果

以上是一种获取 gitee 上配置的规则，还有其他规则如下。

```
// 获取配置规则：根据前缀匹配
/{application-name}/{profiles}[/{label}]
/{application-name}-{profiles}.properties
[/{label}]/{application-name}-{profiles}.properties

application-name // 服务名称
profile // 环境名称：开发、测试、预生产、生产，分别对应于 dev、qa、pre、prd
lable // 仓库分支、默认 master 分支

// 匹配原则：从前缀开始。如 properties 格式，均可换成 yml、json。大家可以实践一下
```

至此，从 gitee 上读取配置的服务端就搭建完成了。下面来看业务服务，学习如何从服务端获取配置。

5.3.3　创建配置的客户端

1．准备工作

创建一个 Spring Boot 的 Web 项目 cloud-config-client，并且将项目注册到 Eureka 注册中心，注册成功后的界面如图 5-5 所示。

图 5-5　cloud-config-client 注册成功

2．配置配置中心客户端

在 pom 文件中添加以下依赖。

```xml
<dependencies>
    <!-- web -->
    <dependency>
        <groupId>org.springframework.boot</groupId>
        <artifactId>spring-boot-starter-web</artifactId>
    </dependency>

    <!-- eureka 客户端 -->
    <dependency>
        <groupId>org.springframework.cloud</groupId>
        <artifactId>spring-cloud-starter-netflix-eureka-client</artifactId>
            <version>2.1.2.RELEASE</version>
    </dependency>


    <!-- 测试 -->
    <dependency>
        <groupId>org.springframework.boot</groupId>
<artifactId>spring-boot-starter-test</artifactId>
        <scope>test</scope>
    </dependency>

    <!-- 配置中心客户端：config-client -->
    <dependency>
        <groupId>org.springframework.cloud</groupId>
        <artifactId>spring-cloud-config-client</artifactId>
    </dependency>

    <dependency>
        <groupId>org.springframework.cloud</groupId>
        <artifactId>spring-cloud-starter-bootstrap</artifactId>
    </dependency>
</dependencies>
```

在配置文件中添加了配置中心客户端的依赖 org.springframework.cloud:spring-cloud-config-client。由于使用的 Spring Cloud 版本是 2020.0.2，还需要添加 org.springframework.cloud:spring-cloud-starter-bootstrap 这个 jar 包，用于后面读取 bootstrap.yml 文件。

此时的项目中不用 application.yml，而用 bootstrap.yml。bootstrap.yml 文件的内容如下。

```yml
#服务端口
server:
  port: 8011

spring:
  application:
    name: config-client

#先注释配置中心
  cloud:
    config:
      discovery:
        #表示使用 discovery 中的 config-server，而不是自己指定 uri，默认 false
        enabled: true
        service-id: config-server-git
      profile: dev
      label: master
#注册中心-此时注册中用于找到 config-server
eureka:
  client:
    #设置服务注册中心的 URL
    service-url:
      defaultZone: http://peer1:8761/eureka/
  instance:
    #服务刷新时间配置，每隔这个时间会主动心跳一次
    lease-renewal-interval-in-seconds: 1
    #服务过期时间配置，超过这个时间没有接收到心跳，Eureka Server 就会将这个实例剔除
    lease-expiration-duration-in-seconds: 1
    instance-id: config-client:${server.port}
```

以下部分是与配置中心服务端相关的配置。

```yml
spring:
  application:
```

```
    name: config-client
#先注释配置中心
cloud:
  config:
    discovery:
      #表示使用 discovery 中的 config server，而不是自己指定 uri，默认 false
      enabled: true
      service-id: config-server-git
    profile: dev
    label: master
```

- ☑ spring.cloud.config.discovery.enabled: true：表示开启配置中心。
- ☑ spring.cloud.config.discovery.service-id:config-server-git：表示配置中心服务端在注册中心的名称。
- ☑ spring.cloud.config.profile:dev：表示这里配置的是在 Git 上创建的文件中最后一个"-"后面的内容，用于区分环境。
- ☑ spring.cloud.config.label:master：表示配置对应在 Git 上的分支。

3．读取配置中心的配置

在 cloud-config-client 项目中重新编写一个 Controller，将 gitee 上配置的 key 值注入属性，返回给客户端调用。客户端通过请求此接口，可以拿到配置中心的配置。

```
@RestController
@RequestMapping("/config")
public class ConfigController {

    @Value("${testKey}")
    private String testKey;

    @GetMapping("/testKey")
    public String testKey(){
        return testKey;
    }
}
```

访问 http://localhost:8011/config/testKey，配置中心客户端请求成功，如图 5-6 所示。

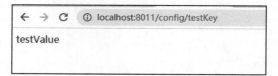

图 5-6　配置中心客户端请求成功

5.3.4　配置的手动刷新

在 5.3.3 节的例子中，如果在 gitee 上修改了配置内容，再访问配置中心的服务端，可以发现修改的内容会实时更新，而在配置中心的客户端，却没有获取到最新值。

下面用 config-client-qa.yml 文件来举例。我们需要修改 bootstrap.yml，将 profile 改成 qa。

```
spring.cloud.config.profile=qa
```

测试一下。将 gitee 上 config-client-qa.yml 中的配置 testKey: test-env 修改成 testKey: test-env-new。

请求配置中心服务端 http://localhost:6001/master/config-client-qa.yml，得到的配置内容如下。

```
testKey: test-env-new
```

请求配置中心的客户端获取配置的接口 http://localhost:8011/config/testKey，得到的依然是旧的数据。

```
test-env
```

如果想让配置中心的客户端感知到 gitee 上的变化，需要改造一下 cloud-config-client 项目，在它的 Controller 上添加一个注解@RefreshScope。代码如下。

```
@RefreshScope
@RestController
@RequestMapping("/config")
public class ConfigController {

    @Value("${testKey}")
    private String testKey;

    @GetMapping("/testKey")
    public String testKey(){
```

```
        return testKey;
    }

}
```

在 pom 中添加如下依赖。

```xml
<dependency>
    <groupId>org.springframework.boot</groupId>
    <artifactId>spring-boot-starter-actuator</artifactId>
</dependency>
```

修改 bootstrap.yml，开启刷新端点。

```yml
management:
  endpoints:
    web:
      exposure:
        #yml 加双引号，properties 不用加
        include: "*"
```

当在 gitee 上修改文件时，在 Postman 上向 cloud-config-client 发送一个 post 请求。

http://localhost:8011/actuator/refresh

再请求配置中心的客户端获取配置的接口 http://localhost:8011/config/testKey，返回的数据就变成 gitee 上更新的数据了。

5.3.5 配置的自动刷新

5.3.4 节讲了手动刷新配置，在 gitee 上修改配置后，需要手动去更新每个配置中心的客户端。但这样对于有成千上万的微服务的大型项目来说是不方便的。Spring Cloud 也支持自动刷新，配置中心自动刷新原理如图 5-7 所示。

自动刷新的流程如下。

（1）gitee 上的修改会触发钩子函数，钩子函数请求一个配置中心服务端的地址 URL-A（这个地址在 gitee 的管理界面可以设置）。

（2）URL-A 调用服务的端点 actuator/bus-refresh。

（3）通过步骤（2），将配置更新的消息发送给消息队列 RabbitMQ。

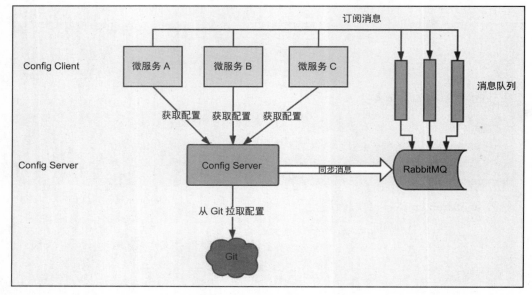

图 5-7　配置中心自动刷新原理

（4）其他所有定义 RabbitMQ 的服务监听到消息后，更新自己的配置。

配置的自动刷新具体实现步骤如下，需要在每个配置中心的客户端都做修改。

（1）在 pom 中引入下面依赖。

```xml
<!-- 服务监控开启refresh 端口 -->
<dependency>
    <groupId>org.springframework.boot</groupId>
    <artifactId>spring-boot-starter-actuator</artifactId>
</dependency>
<!--总线 -->
<dependency>
    <groupId>org.springframework.cloud</groupId>
    <artifactId>spring-cloud-starter-bus-amqp</artifactId>
</dependency>
```

（2）修改 application.yml。

```yaml
spring:
  rabbitmq:
    host: localhost
    port: 5672
```

```
    username: guest
    password: guest
```

（3）开发接收钩子函数调用的接口。

```
@RestController
@RequestMapping("/git-webhook")
public class WebhookController {

    @Autowired
    private RestTemplate restTemplate;

    @PostMapping("/bus-refresh")
    public String refresh() {

        HttpHeaders httpHeaders = new HttpHeaders();
        httpHeaders.add(HttpHeaders.CONTENT_TYPE,"application/json");
        HttpEntity<String> request = new HttpEntity<>(httpHeaders);
        ResponseEntity<String> stringResponseEntity = restTemplate.postForEntity("http://localhost:8011/actuator/bus-refresh", request, String.class);

        return "webhook 刷新成功";
    }
}
```

（4）设置钩子函数，WebHooks 设置如图 5-8 所示。

图 5-8　WebHooks 设置

大厂面试

面试官：在生产环境中，你是如何对配置进行自动刷新的？

回答分析：通过上面的学习，这个问题不难回答。但如果直接说这个答案，是有一些风险的。在线上环境中，每修改一个配置，如果操作人员失误将配置写错，而用到这个配置的全部微服务又都进行了这个错误配置的自动更新，这带来的影响将会是灾难性的。

所以应该通过手动更新来做小范围验证，如果验证正确，再进行大规模配置的更新。回答这个问题时，先把解决方案告诉面试官，然后再带上自己对于安全操作的思考，以及实际落地的方案。这会让面试官对你高看一眼。

5.3.6 在 MySQL 上创建配置

如果读者觉得在 gitee 上配置不方便的话，Spring Cloud Config 也支持将配置存在 MySQL 中。下面来简单介绍一下操作步骤。

（1）创建 MySQL 的数据库和表。

创建一个数据库 config。在 config 中创建一张表 config_server。建表语句如下。

```
CREATE TABLE `config_server` (
 `config_key` varchar(128) DEFAULT NULL,
 `config_value` varchar(128) DEFAULT NULL,
 `application_name` varchar(128) DEFAULT NULL,
 `config_profile` varchar(128) DEFAULT NULL,
 `config_label` varchar(128) DEFAULT NULL
) ENGINE=InnoDB DEFAULT CHARSET=utf8;
```

config_server 表中各字段的意义如下。

- ☑ config_key：配置项目的 key 值。
- ☑ config_value：配置项目的 value 值。
- ☑ application_name：应用名。
- ☑ config_profile：环境名，使用前面介绍的 dev、qa、pre、prd。
- ☑ config_label：相当于 Git 上的分支。

（2）在 pom 中引入数据库依赖。

```xml
<!--数据库-->
<!--连接 MySQL 数据库相关 jar 包-->
<dependency>
    <groupId>org.springframework.boot</groupId>
    <artifactId>spring-boot-starter-jdbc</artifactId>
</dependency>
<dependency>
    <groupId>mysql</groupId>
    <artifactId>mysql-connector-java</artifactId>
    <version>8.0.13</version>
</dependency>
```

（3）修改 bootstrap.yml。

```yaml
spring:
  application:
    name: config-server-git
  cloud:
    config:
      server:
        default-label: qa
        jdbc:
          sql: SELECT config_key , config_value FROM config_server where application_name=? and config_profile=? and config_label=?
  datasource:
    driver-class-name: com.mysql.cj.jdbc.Driver
    url: jdbc:mysql://localhost:3306/config?characterEncoding=UTF-8&serverTimezone=Asia/Shanghai
    username: root
    password: root
  profiles:
    active: jdbc
```

这样就脱离了 gitee，可以将配置信息的 key-value 值插入数据库中了。

5.3.7 配置内容对称加密

Spring Cloud Config 在 gitee 上默认是明文存储的，在生产中会有一些敏感数据，如数据库的用户名和密码等，明文存储是不安全的，需要对它们进行加密存储。下面我们来学习如何进行配置加密。加密分为对称加密和非对称加密，如果可以使用同一个密钥对信息进行加密和解密，就是对称加密。

1. 安装 JCE

配置中心服务端的加密是依赖于 JCE（Java cryptography extension，Java 加密扩展）的，所以在使用加密功能前，需要安装 JCE。首先下载 JCE 包，地址是 https://www.oracle.com/java/technologies/javase-jce8-downloads.html，如图 5-9 所示。

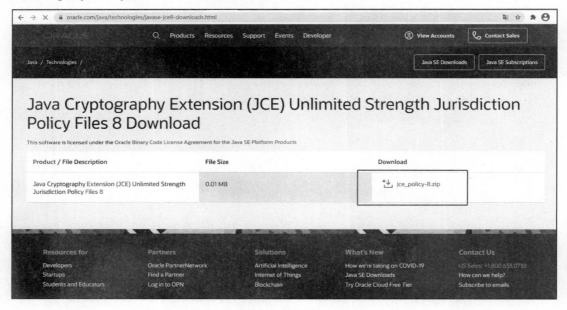

图 5-9　下载 JCE

解压下载的 JCE 包后有两个文件，即 local_policy.jar 和 US_export_policy.jar。用这两个文件将 jre\lib\security 目录下的两个文件替换。

至此，JCE 就安装完成了。

2. 配置加密密钥

修改 application.yml，配置密钥。

```
encrypt:
  key: cpf12345678
```

启动配置中心服务端后，会暴露 encrypt 和 decrypt 两个端点。测试一下加密和解密的效果，加密的效果如图 5-10 所示。

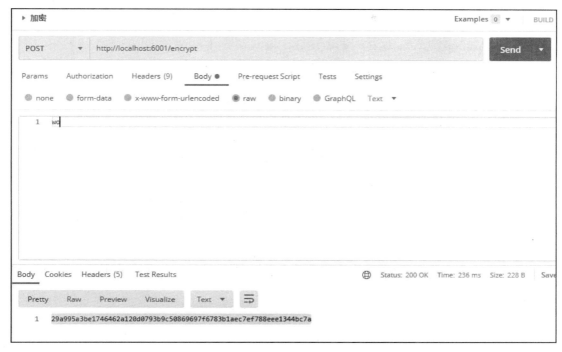

图 5-10　加密效果

解密的效果如图 5-11 所示。

3．修改 gitee 上的配置

将原来的配置改成如下形式。

```
testKey:
'{cipher}29a995a3be1746462a120d0793b9c50869697f6783b1aec7ef788eee1344bc7a'
```

在配置前面加{cipher}，表明这是一个加密内容。后面的 29a995a3be1746462a120d0793b9c50869697f6783b1aec7ef788eee1344bc7a 就是通过 encrypt 加密后的值。

4．校验配置中心服务端的结果

访问 http://localhost:6001/master/config-client-qa.yml。解密结果如图 5-12 所示。

这样，配置中心的客户端就能使用正常的配置了。

图 5-11　解密效果

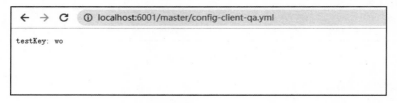

图 5-12　配置中心解密结果

5.3.8　配置内容非对称加密

5.3.7 节讲的是对配置内容的对称加密，一旦密钥泄密，这种加密方式还是不安全的。Spring Cloud Config 还支持非对称加密。非对称加密算法需要两个密钥，即公开密钥（public key）和私有密钥（private key）。公开密钥与私有密钥是一对，如果用公开密钥对数据进行加密，只有用对应的私有密钥才能解密；如果用私有密钥对数据进行加密，那么只有用对应的公开密钥才能解密。因为加密和解密使用的是两个不同的密钥，所以这种算法叫作非对称加密算法。

利用 JDK 生成一个密钥对，公钥用于加密，私钥用于解密。但在非对称加密提高安全性的同时，效率也会降低。下面来实践一下。

（1）首先生成 keystore。

进入 Java 的安装目录，执行以下命令。

```
PS D:\Java\jdk1.8.0_131\bin> .\keytool.exe -genkeypair -alias config-
server-key -keyalg RSA -keystore e:\server.jks
输入密钥库口令：
再次输入新口令：
您的名字与姓氏是什么？
  [Unknown]:  chao
您的组织单位名称是什么？
  [Unknown]:  msb
您的组织名称是什么？
  [Unknown]:  msb
您所在的城市或区域名称是什么？
  [Unknown]:  beijing
您所在的省/市/自治区名称是什么？
  [Unknown]:  beijing
该单位的双字母国家/地区代码是什么？
  [Unknown]:  cn
CN=chao, OU=msb, O=msb, L=beijing, ST=beijing, C=cn 是否正确？
  [否]:  y

输入 <config-server-key> 的密钥口令
        (如果和密钥库口令相同, 按 Enter 键):
再次输入新口令：
PS D:\Java\jdk1.8.0_131\bin>
```

命令执行后，就在 E 盘下生成了一个 server.jks 文件。

（2）下面配置 config server。

将上一步生成的 server.jks 文件复制到 resources 目录下。同时修改 application.yml，增加如下配置。

```
encrypt:
  key-store:
    location: classpath:server.jks
    password: ku123456
```

```
alias: config-server-key
secret: miyue123456
```

以上配置中各个参数的含义如下。
- ☑ location：jks 文件路径。
- ☑ password：密钥库的密码。
- ☑ alias：上面命令中设置的别名。
- ☑ secret：密钥的密码。

（3）最后进行测试。

通过 Postman 测试一下加密和解密功能，如图 5-13 和图 5-14 所示。

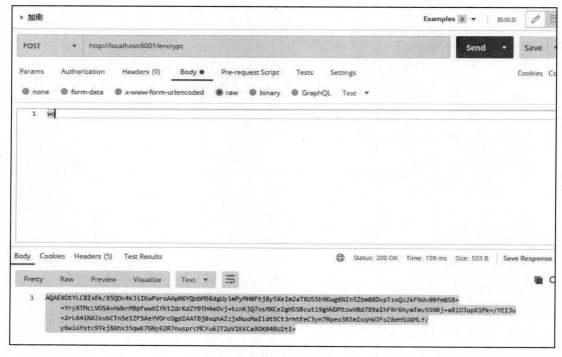

图 5-13　非对称加密

修改 gitee 上 config-client-qa.yml 中的配置内容如下。

```
testKey: '{cipher}AQAEXOtYLCBIxEk/X5QOc4kJiIKwPoroAApN6YQobM96dgUylmPyM
H0FtjBy5XeIm2aTXUS5h9Kwg6NIn5Zbm0BDvpTsxQz2kFNdv08fm8SB++YryXTMcLVOSAvhW
krM9pFwwVZfKtZdrKdZY9TH4eOvj+tcnKjQ7xsMXCe2gHSSBcut19gVWDMtswV0d7B9aIhFR
```

```
r8hymTmv55NRj+aBlOJupX1Pk+/YEI3u+2rL64iNAJsvbCTn5e1ZF5AeYVOrcOgdIAAT8j0x
qhAZzjxNuoMaIldt5Ct3rhtEeC3yn7Rpeo38ImIxqhWJFoZ8eHSUXMLf/y6wioYstc9Tkj8X
hx35qwK7GNy62R7nusprcMCYu6JT2pV1KKCa9OXB4BU2tI='
```

图 5-14 非对称解密

启动服务，访问配置中心服务端，结果如图 5-15 所示。

图 5-15 访问配置中心服务端结果

5.3.9 配置中心安全认证

前面实践过的配置中心，都是可以任意访问的。如果项目上线，这种访问方式是非常不安全的。所以本节学习对配置中心添加认证信息。首先修改 cloud-config-server-git 项目。

（1）在 pom 中添加如下依赖。

```xml
<dependency>
    <groupId>org.springframework.boot</groupId>
    <artifactId>spring-boot-starter-security</artifactId>
</dependency>
```

（2）修改 application.yml。

```yaml
spring:
  security:
    user:
      name: root
      password: root
```

再次访问 http://localhost:6001/master/config-client-qa.yml，发现配置中心添加认证后需要输入用户名和密码了，如图 5-16 所示。输入正确的用户名和密码后，才可以看到正确的配置。

图 5-16　配置中心添加认证

5.3.10　高可用配置中心

高可用配置中心，其实就是搭建多个 config-server 的服务，保证其在 Eureka 中的服务名一致即可。因为在配置中心客户端使用的时候，只认服务名。

5.4　Nacos 配置中心使用

在 2.4 节中讲过 Nacos 作为注册中心的用法，下面介绍 Nacos 作为配置中心的用法。

5.4.1 Nacos 配置中心的基本使用

（1）Nacos 配置中心服务端。

启动 Nacos，就相当于启动了配置中心的存储和配置中心服务端。单击左侧的"配置管理"，新建一个配置，如图 5-17 所示。

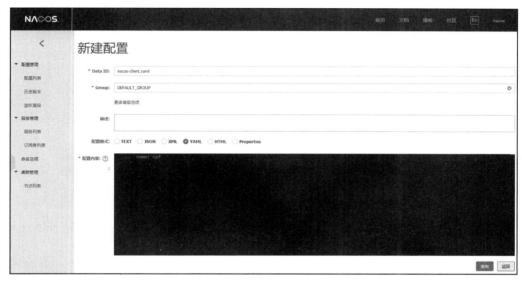

图 5-17 Nacos 创建配置

创建完成后的配置列表如图 5-18 所示。

图 5-18 配置列表

（2）创建 Nacos 客户端项目 nacos-client。

修改 pom 文件，添加如下代码。

```xml
<dependency>
    <groupId>com.alibaba.cloud</groupId>
    <artifactId>spring-cloud-starter-alibaba-nacos-config</artifactId>
    <version>2.2.3.RELEASE</version>
</dependency>
```

（3）修改 bootstrap.yml。

```yaml
spring:
  application:
    name: nacos-client
  cloud:
    nacos:
      config:
        server-addr: 127.0.0.1:8848
        file-extension: yaml
        group: DEFAULT_GROUP
```

在 Nacos 中，默认新建配置的 dataId 对应 ${spring.application.name}.${spring.cloud.nacos.config.file-extension=yaml}。

spring.cloud.nacos.config.server-addr 对应配置中心的 IP 和端口，如果是集群用","分割。

（4）编写代码并进行测试。

提供一个接口，用来返回 Nacos 中的配置。

```java
@SpringBootApplication
@RestController
@RefreshScope
public class ApiDemoApplication {

    public static void main(String[] args) {
        SpringApplication.run(ApiDemoApplication.class, args);
    }

    @Value("${name}")
    private String name;
```

```
@GetMapping("/test-nacos-client")
public String testNacosClient(){

    return "name:"+name;
}
```

访问 http://localhost:8086/test-nacos-client，即可得到对应的配置 cpf。

5.4.2 Nacos 配置扩展

（1）自定义前缀。

```
spring:
  application:
    name: nacos-client
  cloud:
    nacos:
      config:
        server-addr: 127.0.0.1:8848
        file-extension: yaml
        group: DEFAULT_GROUP
        prefix: test
```

该配置对应 Nacos 配置列表中的值，如图 5-19 所示。

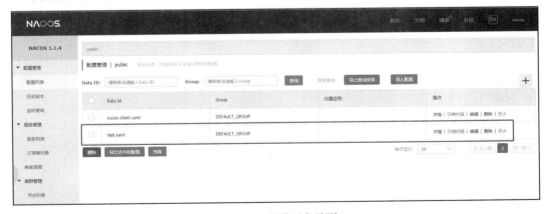

图 5-19 配置列表界面

（2）配置命名空间。

前面的配置都没有配置命名空间，默认是 public 命名空间。新建命名空间 dev，如图 5-20 所示。

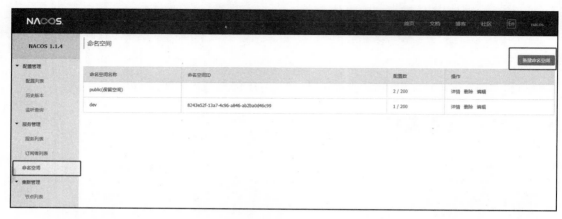

图 5-20　新建命名空间

在 dev 命名空间下创建一个配置，如图 5-21 所示。

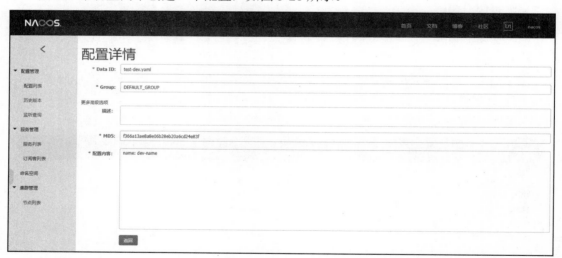

图 5-21　在 dev 下创建配置

修改 bootstrap.yml。

```
spring:
  application:
```

```
    name: nacos-client
cloud:
  nacos:
    config:
      server-addr: 127.0.0.1:8848
      file-extension: yaml
      group: DEFAULT_GROUP
      prefix: test-dev
      namespace: 8243e52f-13a7-4c96-a846-ab2ba0d46c99
```

上述配置中的 spring.cloud.nacos.config.namespace 对应于 dev 的命名空间 ID。

5.4.3 Nacos 模型管理

在实际生产中，通过 namespace 来区分不同的环境，如 dev 表示开发环境、qa 表示测试环境、pre 表示预生产环境、prd 表示生产环境。然后用 group 来区分不同的项目，如开发环境中有用户服务、订单服务，都是通过设置不同的 group 来实现的。group 在创建配置文件时指定，dataId 就是具体的配置文件名称。

5.5 小　　结

本章从配置中心的应用场景出发，分析了配置中心的设计思路，了解了设计思路之后，其实就具备了做一个配置中心的设计能力。然后讲解了 Spring Cloud 中主流的配置中心 Config 和 Nacos 的用法，以及可以在 Config 中进行配置的对称加密和非对称加密。在 Nacos 中讲解了命名空间的创建和生产中如何管理项目的配置。通过本章的学习基本可以达到在生产环境中使用配置中心的能力。

第 6 章

服 务 网 关

在微服务项目中，当项目很大时，微服务有成百上千个。不同的微服务一般都会有不同的地址，而一个功能往往需要调用多个服务才能完成，如果让调用方直接与各个微服务通信，会产生如下几个问题。

- ☑ 客户端多次请求不同的微服务，增加了客户端的复杂性。
- ☑ 认证复杂，因为每个服务都需要给请求做独立认证。
- ☑ 难以重构，随着需求的变更，项目重构在所难免，这其中有可能需要进行服务的合并与拆分，如果客户端直接和服务端通信，重构将很难实施，因为后端的重构会影响到客户端的调用方式。

以上这些问题的解决办法很简单，其实就是在请求和服务之间做一层解耦，多一个中间层，中间层统一接收用户的请求，做一些通用的功能，如权限认证等。而这个中间层就是网关。Spring Cloud 通过 Gateway 组件，提供了网关功能。

6.1 网关 Gateway 的基本使用

本节将搭建如下 3 个组件。

- ☑ passenger-api：模拟一个微服务。
- ☑ cloud-gateway：模拟一个网关。
- ☑ cloud-eureka：注册中心。

首先创建一个 Eureka 服务端，并启动 Eureka（参考 2.3 节中 Eureka 服务端的搭建）。

6.1.1 微服务搭建 passenger-api

基于 Spring Boot 创建一个 Web 应用，创建一个接口 GatewayController，输出"hello gateway I'm passenger-api"。

```
@RestController
@RequestMapping("/gateway")
public class GatewayController {

    @GetMapping("/hello")
    public String hello(){
        return "hello gateway I'm passenger-api";
    }
}
```

设置好此应用的 hostname，否则后面访问的时候解析不了目标地址。

```
server:
  port: 8090
spring:
  application:
    name: passenger-api
eureka:
  client:
    serviceUrl:
      defaultZone: http://localhost:8761/eureka/
  instance:
    hostname: localhost
```

启动 passenger-api，并注册到 cloud-eureka，如图 6-1 所示，passenger-api 注册成功。

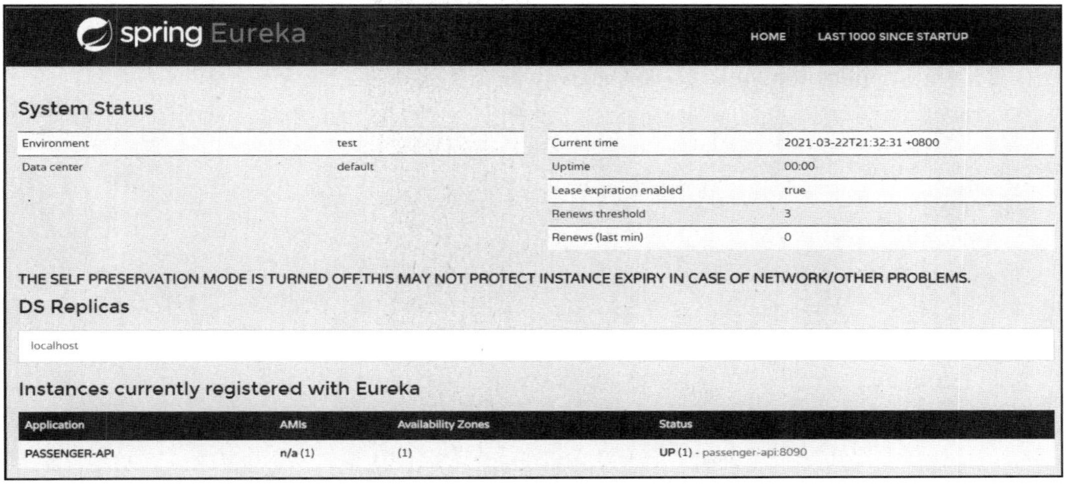

图 6-1　passenger-api 注册成功

6.1.2 Gateway 网关搭建 cloud-gateway

Gateway 网关搭建 cloud-gateway，具体的搭建步骤如下。

（1）创建一个 Spring Boot 项目 cloud-gateway，并添加如下依赖。

```xml
<dependencies>
    <dependency>
        <groupId>org.springframework.boot</groupId>
        <artifactId>spring-boot-starter</artifactId>
    </dependency>

    <dependency>
        <groupId>org.springframework.boot</groupId>
        <artifactId>spring-boot-starter-test</artifactId>
        <scope>test</scope>
    </dependency>

    <dependency>
        <groupId>org.springframework.cloud</groupId>
        <artifactId>spring-cloud-starter-gateway</artifactId>
    </dependency>
</dependencies>
```

（2）修改 application.yml，添加 Gateway 路由配置。

```yaml
server:
  port: 9001
spring:
  application:
    name: cloud-gateway
  cloud:
    gateway:
      routes:
        - id: gateway1
          predicates:
            - Path=/gateway/**
          uri: http://localhost:8090/
```

在上述代码中，id 是路由规则的 ID；predicates 是断言，当请求的地址中有/gateway/时和此条路由规则匹配；uri 是请求的目标地址，将匹配到的请求转发到目标地址。

启动 cloud-gateway，测试一下网关请求，如图 6-2 所示。

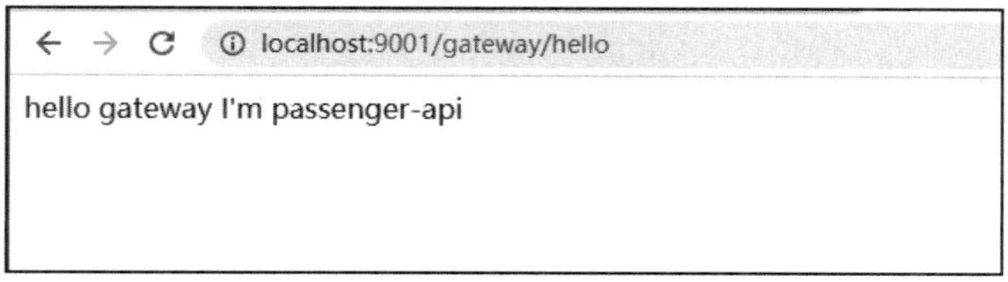

图 6-2　网关请求

当请求 http://localhost:9001/gateway/hello 时，会将请求转发到 http://localhost:8090/gateway/hello。

（3）与 Eureka 结合，修改 cloud-gateway 的 pom，添加 eureka 的依赖。

```xml
<dependency>
    <groupId>org.springframework.cloud</groupId>
    <artifactId>spring-cloud-starter-netflix-eureka-client</artifactId>
    <exclusions>
        <exclusion>
            <groupId>org.springframework.boot</groupId>
            <artifactId>spring-boot-starter-web</artifactId>
        </exclusion>
    </exclusions>
</dependency>
```

此时的依赖要将 org.springframework.boot:spring-boot-starter-web 排除。修改 application.yml。

```yaml
server:
  port: 9001
spring:
  application:
    name: cloud-gateway
  cloud:
    gateway:
      routes:
        - id: gateway1
          predicates:
```

```
        - Path=/gateway/**
      uri: lb://passenger-api
    discovery:
      locator:
        enabled: true
eureka:
  client:
    service-url:
      defaultZone: http://localhost:8761/eureka/
  instance:
    instance-id: ${eureka.instance.hostname}:${server.port}
    hostname: localhost
```

注意此时修改的如下两点。

☑ 将 spring.cloud.gateway.discovery.locator 设置为 true，否则 cloud-gateway 无法从注册中心发现服务。

☑ 配置 spring.cloud.gateway.routes[0].uri=lb://passenger-api/。将原来的 http 改成 lb，后面的 ip：port 改成服务名（虚拟主机名）。

重启服务，进行验证，如图 6-3 所示，请求转发 eureka 服务成功。

图 6-3　请求转发 eureka 服务成功

6.1.3　Java 类加载器层级结构

我们先来看看 Java 类加载的实现原理。在 Java 应用启动时，将会创建如下 3 个类加载器。

（1）Bootstrap ClassLoader。Bootstrap 加载器在虚拟机中用 C++语言编写而成，在 Java 虚拟机启动时初始化，它主要负责加载路径为%JAVA_HOME%/jre/lib 或者通过参数 -Xbootclasspath 指定的路径以及%JAVA_HOME%/jre/classes 中的类。

（2）ExtClassLoader。Bootstrap ClassLoader 创建了 ExtClassLoader，ExtClassLoader 是用 Java 编写的，具体来说就是 sun.misc.Launcher$ExtClassLoader，ExtClassLoader 继承自 URLClassLoader，负责加载%JAVA_HOME%/jre/lib/ext 路径下的扩展类库，以及 java.ext.dirs

系统变量指定的路径中的类。

（3）AppClassLoader。Bootstrap ClassLoader 创建完 ExtClassLoader 之后，就会创建 AppClassLoader，并且将 AppClassLoader 的父加载器指定为 ExtClassLoader。AppClassLoader 也是用 Java 编写成的，它的实现类是 sun.misc.Launcher$AppClassLoader，也是继承自 URLClassLoader，在 ClassLoader 类中有一个方法 getSystemClassLoader，该方法返回的正是 AppClassLoader。AppClassLoader 主要负责加载 java.class.path 所指定位置的类，它也是 Java 程序默认的类加载器。通过 java -cp 命令传入的路径的原理也就在于此。

如图 6-4 所示，在 Java 中类加载器总是按照负责关系排列，通常当类加载器被要求加载特定的类或资源时，它首先将请求委托给父类加载器，然后仅在父类加载器找不到请求的类或资源后，再自己加载，这种加载机制也称之为双亲委派机制。

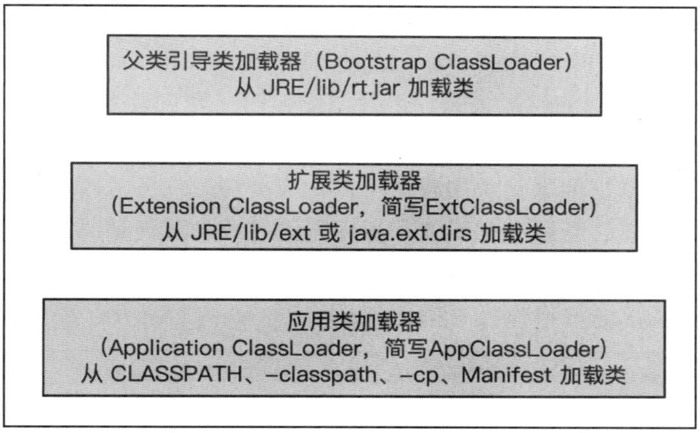

```
父类引导类加载器（Bootstrap ClassLoader）
从 JRE/lib/rt.jar 加载类

扩展类加载器
（Extension ClassLoader，简写 ExtClassLoader）
从 JRE/lib/ext 或 java.ext.dirs 加载类

应用类加载器
（Application ClassLoader，简写 AppClassLoader）
从 CLASSPATH、–classpath、–cp、Manifest 加载类
```

图 6-4　JDK 类加载器体系

下面来看一段 Java 代码。

```java
public class Demo{
    public static void main(String[] arg){
        ClassLoader c = Demo.class.getClassLoader();// 获取 Demo 类应用类加载器
        System.out.println(c.getClass().getSimpleName());
        ClassLoader c1 = c.getParent();              // 获取 c 类加载器的父类加载器
        System.out.println(c1.getClass().getSimpleName());
        ClassLoader c2 = c1.getParent();             // 获取 c1 类加载器的父类加载器
        System.out.println(c2);
    }
}
```

上述代码输出结果如下。

```
AppClassLoader
ExtClassLoader
null
```

通过以上代码和输出结果可以看出 Demo 类是由 AppClassLoader 加载的，AppClassLoader 的父加载器是 ExtClassLoader，但是 ExtClassLoader 的父加载器为 null，前面说到 Bootstrap ClassLoader 是用 C++语言编写的，在逻辑上并不存在 Bootstrap ClassLoader 的类实体，所以在 Java 程序代码中试图打印时，就会看到输出为 null。

6.1.4　Java 双亲委派机制原理

6.1.3 节中提到了 URLClassLoader、AppClassLoader 和 ExtClassLoader，下面来看看 ClassLoader 类和它们之间的层级关系，以及源码如何实现双亲委派机制。

通俗来讲，双亲委派机制就是某个类加载器在接到加载类的请求时，会先将加载类的任务委托给父类加载器，如果父类加载器可以完成类加载任务，就成功返回，否则一直委托到最顶层的类加载器。当所有父类加载器均无法完成加载任务时，才会自己加载。先来看 ExtClassLoader 的继承体系，如图 6-5 所示，其继承自 URLClassLoader，而 URLClassLoader 继承自 SecureClassLoader，根类为 ClassLoader。

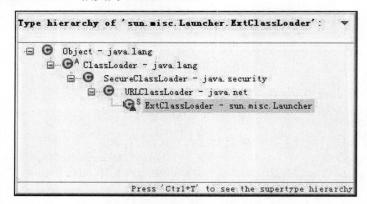

图 6-5　ExtClassLoader 的继承体系

然后来看 AppClassLoader 的继承体系，如图 6-6 所示，它也是继承自 URLClassLoader，而 URLClassLoader 继承自 SecureClassLoader，根类为 ClassLoader 类。

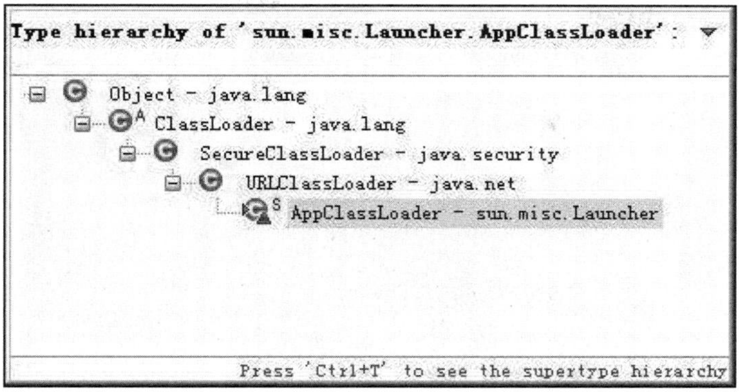

图 6-6　AppClassLoader 的继承体系

通过图 6-5 和图 6-6 可以看出，类加载器均继承自 java.lang.ClassLoader 抽象类。

6.1.5　Java ClassLoader 类的原理

下面直接来看 ClassLoader 类的核心加载方法 loadClass 的实现过程，通常使用该方法来加载需要的类，而该方法也实现了双亲委派机制。

```java
public abstract class ClassLoader {
    // name 为类全限定名，resolve 指定是否加载类后完成解析
    protected Class<?> loadClass(String name, boolean resolve) throws ClassNotFoundException{
        // 获取锁，避免多线程对同一个类进行加载
        synchronized (getClassLoadingLock(name)) {
            // 首先看看当前类是否已经被加载
            Class<?> c = findLoadedClass(name);
            if (c == null) {
                try {
                    // 如果指定了父加载器，那么尝试从父加载加载
                    if (parent != null) {
                        c = parent.loadClass(name, false);
                    } else {
                        // 否则尝试通过 Bootstrap 类加载器加载
                        c = findBootstrapClassOrNull(name);
                    }
                } catch (ClassNotFoundException e) {
```

```
        }
        // 如果仍未找到，那么调用子类重写的 findClass 方法查找类
        if (c == null) {
            long t1 = System.nanoTime();
            c = findClass(name);
        }
    }
    // 如果指定了解析类，那么进行解析
    if (resolve) {
        resolveClass(c);
    }
    return c;
    }
}
```

从上述代码中看到，loadClass 方法实现了查询已加载类的功能。同时如果指定了 parent 父类加载器，那么将会实现双亲委派机制来完成加载；如果没有指定，那么在自己加载之前还是会通过 Bootstrap 加载器尝试加载；如果还是无法加载，那么调用 findClass 方法尝试自己加载。现在来看 findClass 的默认实现代码。

```
protected Class<?> findClass(String name) throws ClassNotFoundException {
    throw new ClassNotFoundException(name);
}
```

从上述代码中可以看到，findClass 默认实现是抛出 ClassNotFoundException，通常如果希望保持双亲委派机制，那么应该实现该方法完成自己的加载逻辑，当然可以直接重写 loadClass 方法来实现该业务逻辑，但这时将会打破双亲委派机制。

6.1.6　Java URLClassLoader 类的原理

下面来看 URLClassLoader 类加载器是如何通过实现 findClass 方法来完成自己的类加载逻辑的。注意，URLClassLoader 类加载器没有重写 loadClass 方法，只是扩展了 findClass 方法实现了自己的加载逻辑。而对于 SecureClassLoader，笔者这里将其省略，因为它里面只定义了加载安全校验的逻辑，主要与 Java 安全管理器交互，所以为了避免影响读者的学习主线，这里将其忽略。

```java
public class URLClassLoader extends SecureClassLoader implements Closeable {
    // 指定查找类的路径
    private final URLClassPath ucp;

    // urls 指定查找的路径链接,parent 指定类加载器的父类
    public URLClassLoader(URL[] urls, ClassLoader parent) {
        // 调用父类构造器设置父加载器
        super(parent);
        // 创建查找类的路径
        SecurityManager security = System.getSecurityManager();
        if (security != null) {
            security.checkCreateClassLoader();
        }
        this.acc = AccessController.getContext();
        ucp = new URLClassPath(urls, acc);
    }

    // 实现自己的类查找逻辑
    protected Class<?> findClass(final String name) throws
ClassNotFoundException{
    final Class<?> result;
    try {
        result = AccessController.doPrivileged(
            new PrivilegedExceptionAction<Class<?>>() {
                public Class<?> run() throws ClassNotFoundException {
                    // 构建类查找路径,并通过 URLClassPath 查找
                    String path = name.replace('.', '/').concat(".class");
                    Resource res = ucp.getResource(path, false);
                    // 找到类,那么调用 defineClass 方法加载类到 JVM 中
                    if (res != null) {
                        try {
                            return defineClass(name, res);
                        } catch (IOException e) {
                            throw new ClassNotFoundException(name, e);
                        }
                    } else {
                        return null;
```

```
            }
          }
        }, acc);
    } catch (java.security.PrivilegedActionException pae) {
        throw (ClassNotFoundException) pae.getException();
    }
    if (result == null) {
        throw new ClassNotFoundException(name);
    }
    return result;
}
```

从上述代码中可以看到，URLClassLoader 就是定义了可以通过指定的 URL 链接寻找类的逻辑，通过扩展 ClassLoader 的 findClass 方法完成了自己的类的加载方法，最终调用了 ClassLoader 的 defineClass 方法完成了类加载。那么现在可以来看看 ExtClassLoader 的源码实现。

```
static class ExtClassLoader extends URLClassLoader {
    // 指定 Ext 扩展类所在的目录信息
    public ExtClassLoader(File[] dirs) throws IOException {
        // 调用 URLClassLoader 类保存加载 URL 信息
        super(getExtURLs(dirs), null, factory);
    }

    // 根据 dirs File 目录信息生成 URL 数组
    private static URL[] getExtURLs(File[] dirs) throws IOException {
        Vector<URL> urls = new Vector<URL>();
        // 遍历目录，将 File 对象转变为 URL 对象并返回
        for (int i = 0; i < dirs.length; i++) {
            String[] files = dirs[i].list();
            if (files != null) {
                for (int j = 0; j < files.length; j++) {
                    if (!files[j].equals("meta-index")) {
                        File f = new File(dirs[i], files[j]);
                        urls.add(getFileURL(f));
                    }
```

```
            }
        }
    }
    URL[] ua = new URL[urls.size()];
    urls.copyInto(ua);
    return ua;
}
```

从上述代码中可以看到，ExtClassLoader 继承自 URLClassLoader，通过将类文件目录转变为 URL 对象传递给 URLClassLoader，还是通过 URLClassLoader 的 findClass 方法来加载类。下面来看 AppClassLoader 的类实现原理。

```
static class AppClassLoader extends URLClassLoader {
    // 直接传递 urls 信息
    AppClassLoader(URL[] urls, ClassLoader parent) {
        super(urls, parent, factory);
    }
}
```

从上述代码中可以看到，AppClassLoader 的实现更为简单，直接传递 URL 信息给 URLClassLoader，最终也是通过 findClass 方法传递进去的 URL 信息查找类并定义类信息。

6.1.7　Java 双亲委派机制的打破

通过上面的源码分析，可以对 JVM 采用的双亲委派类加载机制有了更深的认识。那么接下来考虑一个问题，如何打破双亲委派机制？可能读者会问，为什么要打破这种机制？双亲委派机制的出现是为了防止用户定义和 Bootstrap 类加载器一样的类名，然后又在其中写入了危害 JVM 的代码，如果没有双亲委派机制，那么就可以直接加载这个危险的类，从而造成破坏。但是如果有双亲委派机制，由于这个类必须由 Bootstrap 类加载器来加载，那么当传递到父类加载器时就会返回正确的类实例，这样就不会造成 JVM 的危害类加入虚拟机。

那么为什么又要打破这种机制呢？是这样的，Java 提供了 SPI（service provider interface，服务提供者接口），允许第三方为这些接口提供实现。常见的 SPI 有 JDBC、JCE、JNDI、JAXP 和 JBI 等。这些 SPI 的接口由 Java 核心库来提供，如 JAXP 的 SPI 接口定义包含在 javax.xml.parsers 包中。这些 SPI 的实现代码很可能是作为 Java 应用所依赖的 jar 包被包含进来，可以通过类路径（CLASSPATH）来找到，如实现了 JAXP 的 SPI 的 Apache Xerces

所包含的 jar 包。SPI 接口中的代码经常需要加载具体的实现类。如 JAXP 中的 javax.xml.parsers.DocumentBuilderFactory 类中的 newInstance()方法用来生成一个新的 DocumentBuilderFactory 的实例，这里的实例继承自 javax.xml.parsers.DocumentBuilderFactory。如在 Apache Xerces 中，实现的类是 org.apache.xerces.jaxp.DocumentBuilderFactoryImpl，而问题在于 SPI 的接口是 Java 核心库的一部分，其是由引导类加载器来加载的，SPI 提供商实现的 Java 类由系统类加载器来加载，如果按照双亲委派机制的实现，那么引导类加载器是无法找到 SPI 的实现类的，因为它只加载 Java 的核心库。它也不能代理给系统类加载器，因为它是系统类加载器的顶层类加载器，也就是说，类加载器的双亲委派加载机制无法解决这个问题。

那么如何解决这种问题呢？这时就需要介绍一下从 JDK 1.2 开始引入的线程上下文类加载器（thread context class loader）。可以通过 java.lang.Thread 中的 getContextClassLoader()和 setContextClassLoader (ClassLoader cl)方法来获取和设置线程的上下文类加载器，如果没有调用 setContextClassLoader (ClassLoader cl)方法进行设置线程上下文类加载器的话，那么线程将继承其父线程的上下文类加载器。Java 运行的初始线程的上下文类加载器是系统类加载器，在线程中运行的代码可以通过此类加载器来加载类和资源。线程上下文类加载器正好解决了在 SPI 中调用系统类加载器的问题，如果不做任何的设置，Java 应用的线程的上下文类加载器默认就是系统上下文类加载器，在 SPI 接口的代码中使用线程上下文类加载器，就可以成功地加载到 SPI 实现的类。线程上下文类加载器在很多 SPI 的实现中都会用到。

当然，从源码的分析中也可以看到，可以通过继承 ClassLoader 类，然后将 loadClass 方法重写，这时就完全打破了双亲委派的实现。因为方法都被重写了，自然不存在任何双亲委派的逻辑，但是为了应用程序的安全，一般来说不推荐重写 loadClass 方法，而是通过集成 ClassLoader 重写 findClass 方法，或者可以直接用 URLClassLoader 传递加载路径来完成加载，当然本质上 URLClassLoader 也是通过重写 findClass 方法来加载的。

6.1.8 Java 自定义类加载器

在 ClassLoader 的源码中有两个核心方法，即 loadClass 和 findClass，而且 loadClass 中实现了双亲委派机制。当不指定父加载器 parent 时，也会先到 Bootstrap 中加载，如果没有加载成功，那么将会调用 ClassLoader 的 findClass 方法加载类，且 findClass 方法默认抛出 ClassNotFoundException，同时在 URLClassLoader 类中对 findClass 方法进行了实现，可以通过制定 URL 来加载类文件。下面来看如何自定义类加载器。读者请注意，为了掩饰自定义类加载器的原理，本节首先定义了 3 个类。

```java
public class Common{
    public void test() throws ClassNotFoundException{
        System.out.println("need A class , start contextClassLoader load");
        ClassLoader contextClassLoader = Thread.currentThread().getContextClassLoader();
        contextClassLoader.loadClass("A");
        System.out.println("load success");
    }
}

public class A {
    public void test(B b){
        System.out.println(b);
    }
}

public class B {

}
```

以上代码中的 Common 类的 test 方法需要加载 A 类，A 类的 test 方法需要依赖 B 类对象。首先使用 javac 命令进行编译，然后将代码存到一个独立于项目的文件夹中，供自定义类加载器时使用。

下面讲解 Java 自定义类加载器的两个扩展点。

1. 继承 ClassLoader 重写 loadClass 方法

重写 loadClass 方法时，如果不实现双亲委派机制，那么这种机制将会被破坏。来看代码实现。

```java
// 定义类加载器
static class MyClassLoader extends ClassLoader {
    // 类文件路径
    public File classFileDirectory;

    public MyClassLoader(File classFileDirectory) {
        if (classFileDirectory == null) {
            throw new RuntimeException("目录为null");
```

```java
        }
        if (!classFileDirectory.isDirectory()) {
            throw new RuntimeException("必须是目录");
        }
        this.classFileDirectory = classFileDirectory;
    }
    // 重写loadClass方法
    @Override
    public Class<?> loadClass(String name, boolean resolve) throws ClassNotFoundException {
        Class<?> clazz;
        // 读者注意，这段代码非常重要，如果没有这段代码，将会导致加载失败，原因是找不到
        // java/lang/Object 类，此时重写了 loadClass 相当于打破了双亲委派机制，在Java中不指定
        // 显示继承的类都会默认继承 Object 类，此时由于 Object 是 Bootstrap 类加载器加载的，所以
        // 无法获取它。此时必须调用父类 ClassLoader 加载 Object 类
        try {
            // 保证java/lang/Object 被加载
            clazz = super.loadClass(name, false);
            if (clazz != null) {
                return clazz;
            }
        } catch (Exception e) { // 忽略异常
            // e.printStackTrace();
        }
        try {
            // 获取类文件的 File 对象，使用 FileInputStream 加载后读入 byte 数组，最后
            // 通过 defineClass 方法加载类
            File classFile = findClassFile(name);
            FileInputStream fileInputStream = new FileInputStream(classFile);
            byte[] bytes = new byte[(int) classFile.length()];
            fileInputStream.read(bytes);
            clazz = defineClass(name, bytes, 0, bytes.length);
        } catch (Exception e) {
            throw new ClassNotFoundException();
        }
        return clazz;
    }
```

```java
    // 获取名字为 name 的类文件
    public File findClassFile(String name) {
        // 获取文件夹下的所有文件，遍历找到对应名为 name 的类文件然后返回 File 对象
        File[] files = classFileDirectory.listFiles();
        for (int i = 0; i < files.length; i++) {
            File cur = files[i];
            String fileName = cur.getName();
            // 只加载类文件
            if (fileName.endsWith(".class") && fileName.substring(0, fileName.lastIndexOf(".")).equals(name)) {
                return cur;
            }
        }
        return null;
    }
}
 public static void main(String[] args) throws Exception {
     ClassLoader classLoader = new MyClassLoader(new File("C:\\Users\\hj\\IdeaProjects\\Demo\\class"));
     Class<?> aClazz = classLoader.loadClass("A");
     System.out.println("a1Clazz 类加载器: " + aClazz);
 }
```

上述代码输出结果如下。

```
java    a1Clazz 类加载器: class A
```

成功加载了类，但这种方法不是十分优雅，因为必须依赖 Bootstrap 加载类，而由于在 ClassLoader 中对 Bootstrap 类加载器的使用方法是私有的，所以只能调用 ClassLoader 父类的 loadClass 方法来加载 Object 类。

2. 继承 ClassLoader 重写 findClass 方法

下面来看在实际开发中如何编写自己的类加载器。在保留双亲委派机制的同时实现的类加载器的方法，即重写 findClass 方法。下面是代码实现。

```
package org.com.msb.classloader;
```

```java
import java.io.File;
import java.io.FileInputStream;

/**
 * @author hj
 * @version 1.0
 * @description: TODO
 * @date 2022/3/6 7:50
 */
public class Main1 {
    static class MyClassLoader extends ClassLoader {
        ...
        // 重写ClassLoader的findClass方法
        @Override
        public Class<?> findClass(String name) throws ClassNotFoundException {
            // 此时不需要去手动调用父类的loadClass方法,因为双亲委派没有被打破
            Class<?> clazz;
            try {
                File classFile = findClassFile(name);
                FileInputStream fileInputStream = new FileInputStream(classFile);
                byte[] bytes = new byte[(int) classFile.length()];
                fileInputStream.read(bytes);
                clazz = defineClass(name, bytes, 0, bytes.length);
            } catch (Exception e) {
                throw new ClassNotFoundException();
            }
            return clazz;
        }
        ...
    }

    public static void main(String[] args) throws Exception {
        ClassLoader classLoader = new MyClassLoader(new File("C:\\Users\\hj\\IdeaProjects\\Demo\\class"));
        Class<?> aClazz = classLoader.loadClass("A");
```

```
        System.out.println("a1Clazz 类加载器: " + aClazz);
    }
}
```

上述代码结果输出如下。

```java
    a1Clazz 类加载器: class A
```

在了解了如何自定义类加载器后,来看以下代码发生的情况。

```
public static void main(String[] args) throws Exception {
    // 创建类加载器对象
    ClassLoader classLoader1 = new MyClassLoader(new File("C:\\Users\\hj\\IdeaProjects\\Demo\\class"));
    ClassLoader classLoader2 = new MyClassLoader(new File("C:\\Users\\hj\\IdeaProjects\\Demo\\class"));

    // classLoader1 类加载器加载 A 类
    Class<?> a1Clazz = classLoader1.loadClass("A");
    // classLoader1 类加载器加载 B 类
    Class<?> b1Clazz = classLoader1.loadClass("B");
    // classLoader2 类加载器加载 B 类
    Class<?> b2Clazz = classLoader2.loadClass("B");

    System.out.println("a1Clazz 类加载器: " + a1Clazz.getClassLoader());
    System.out.println("b1Clazz 类加载器: " + b1Clazz.getClassLoader());
    // 创建 A 类和 B 类的对象
    Object aObj1 = a1Clazz.newInstance();
    Object bObj2 = b2Clazz.newInstance();
    // 获取 A 类的 test 方法
    Method aTestMethod = a1Clazz.getDeclaredMethod("test", b1Clazz);
    // 调用 aObj1 对象的 test 方法,传入参数是类加载器 classLoader2 创建的对象 bObj2
    aTestMethod.invoke(aObj1, bObj2);
}
```

上述代码运行结果如下。

```
a1Clazz 类加载器: org.com.msb.classloader.Main$MyClassLoader@6d6f6e28
b1Clazz 类加载器: org.com.msb.classloader.Main$MyClassLoader@6d6f6e28
Exception in thread "main" java.lang.IllegalArgumentException: argument
```

```
type mismatch
    at sun.reflect.NativeMethodAccessorImpl.invoke0(Native Method)
    at sun.reflect.NativeMethodAccessorImpl.invoke
(NativeMethodAccessorImpl.java:62)
    at sun.reflect.DelegatingMethodAccessorImpl.invoke
(DelegatingMethodAccessorImpl.java:43)
    at java.lang.reflect.Method.invoke(Method.java:498)
    at org.com.hj.classloader.Main.main(Main.java:103)
```

上述代码中，用自定义的类加载器创建了 classLoader1 和 classLoader2 两个类加载器对象，然后分别用它们加载了 A 和 B 类，又通过加载 Class 对象创建了 aObj1 和 bObj2 对象。此时通过反射调用 aObj1 方法，由于传入的对象是 classLoader2 的 bObj2 对象，而 bObj2 对象又不是由 classLoader1 加载的，即和 aObj1 对象用的不是同一个类加载器，所以造成了 argument type mismatch 异常。读者在生产环境中如果使用不慎，将会出现这个异常，可以通过原理来进行排错。下面用一张图来描述一下此时创建两个 ClassLoader 对象的类加载器的状态。如图 6-7 所示，此时满足了双亲委派机制，且通过定义的类加载器来进行了类的隔离，因为不同的类加载器之间的类是相互隔离的。

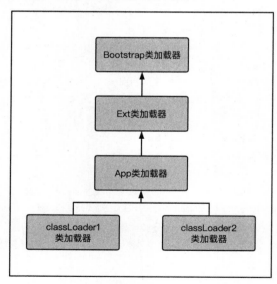

图 6-7　JDK 自定义类加载器结构

继续看以下代码发生的情况。笔者这里通过硬编码的方式，模拟了 SPI 在 Bootstrap 中需要加载系统类加载器才能够加载的类。

```java
static class MyClassLoader extends ClassLoader {
    public File classFileDirectory;
    // 标识当前类加载器是否为父类加载器
    public boolean isParent;

    public MyClassLoader(File classFileDirectory, ClassLoader parent) {
        // 调用 ClassLoader 的构造器, 将 parent 设置为当前类加载器的父类加载器
        super(parent);
        // 如果 parent 为空, 那么当前类加载器就是父类加载
        isParent = parent == null;
        if (classFileDirectory == null) {
            throw new RuntimeException("目录为null");
        }
        if (!classFileDirectory.isDirectory()) {
            throw new RuntimeException("必须是目录");
        }
        this.classFileDirectory = classFileDirectory;
    }

    // 重写 findClass 方法, 和前面一致, 这里省略
    @Override
    public Class<?> findClass(String name) throws ClassNotFoundException {
        ...
    }

    // 获取名字为 name 的类文件
    public File findClassFile(String name) {
        File[] files = classFileDirectory.listFiles();
        for (int i = 0; i < files.length; i++) {
            File cur = files[i];
            String fileName = cur.getName();
            // 只加载类文件
            if (fileName.endsWith(".class") && fileName.substring(0, fileName.lastIndexOf(".")).equals(name)) {
                // Common 类文件只能有父类加载器加载
```

```java
                if (!isParent || name.equals("Common")) {
                    return cur;
                }
            }
        }
        return null;
    }

    public static void main(String[] args) throws Exception {
        // 创建 parent、classLoader1、classLoader2 类加载器, 并将 parent 类加载器设置
为 classLoader1、classLoader2 类加载器的父类加载器
        ClassLoader parent = new MyClassLoader(new File("C:\\Users\\hj\\IdeaProjects\\Demo\\class"), null);
        ClassLoader classLoader1 = new MyClassLoader(new File("C:\\Users\\hj\\IdeaProjects\\Demo\\class"), parent);
        ClassLoader classLoader2 = new MyClassLoader(new File("C:\\Users\\hj\\IdeaProjects\\Demo\\class"), parent);
        // 输出地址信息
        System.out.println("parent 类加载器: " + parent);
        System.out.println("classLoader1 类加载器: " + classLoader1);
        System.out.println("classLoader2 类加载器: " + classLoader2);
        // classLoader1 加载 Common 类, 并调用 test 方法
        Class<?> commonClazz = classLoader1.loadClass("Common");
        System.out.println("commonClazz 类加载器: " + commonClazz.getClassLoader());
        Object commonObj = commonClazz.newInstance();
        Method commonTestMethod = commonClazz.getDeclaredMethod("test");
        commonTestMethod.invoke(commonObj);
    }
}
```

上述代码运行结果如下。

```
parent 类加载器: org.com.msb.classloader.Main$MyClassLoader@6d6f6e28
classLoader1 类加载器: org.com.msb.classloader.Main$MyClassLoader@135fbaa4
```

```
classLoader2 类加载器: org.com.msb.classloader.Main$MyClassLoader@45ee12a7
commonClazz 类加载器: org.com.msb.classloader.Main$MyClassLoader@6d6f6e28
need A class , start contextClassLoader load
Exception in thread "main" java.lang.reflect.InvocationTargetException
    at sun.reflect.NativeMethodAccessorImpl.invoke0(Native Method)
    at sun.reflect.NativeMethodAccessorImpl.invoke
(NativeMethodAccessorImpl.java:62)
    at sun.reflect.DelegatingMethodAccessorImpl.invoke
(DelegatingMethodAccessorImpl.java:43)
    at java.lang.reflect.Method.invoke(Method.java:498)
    at org.com.hj.classloader.Main.main(Main.java:78)
Caused by: java.lang.ClassNotFoundException: A
    at java.net.URLClassLoader.findClass(URLClassLoader.java:381)
    at java.lang.ClassLoader.loadClass(ClassLoader.java:424)
    at sun.misc.Launcher$AppClassLoader.loadClass(Launcher.java:349)
    at java.lang.ClassLoader.loadClass(ClassLoader.java:357)
    at Common.test(Common.java:5)
    ... 5 more
```

Common 类的 test 方法中需要加载类 A，但是由于 Common 类中只能通过线程上下文类加载器加载，而线程上下文类加载器默认是 App 类加载器，App 类加载器自然不知道 A 类在哪里加载，如图 6-8 所示。

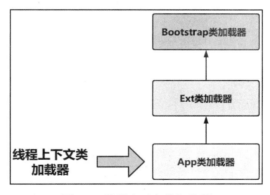

图 6-8　线程上下文类加载器

定义的类加载器的层级结构如图 6-9 所示，classLoader1 和 classLoader2 的父类加载器为 parent 类加载器。

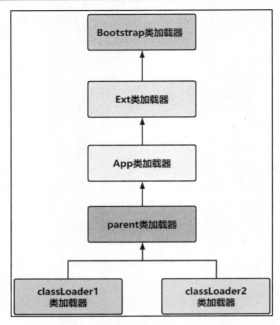

图 6-9　自定义类加载器之父子加载器

从上述代码的运行结果中可以看到，commonClazz 的类加载器为 parent 类加载器，那么这就很形象地表述了需要打破双亲委派机制的原因了，这时父类加载器需要通过子类加载器加载需要的类 A，那么应该怎么做呢？还记得之前说的 setContextClassLoader 方法吗？来看如下代码。

```
ClassLoader parent = new MyClassLoader(new File("C:\\Users\\hj\\
IdeaProjects\\Demo\\class"), null);
ClassLoader classLoader1 = new MyClassLoader(new
 File("C:\\Users\\hj\\IdeaProjects\\Demo\\class"), parent);
ClassLoader classLoader2 = new MyClassLoader(ne
w File("C:\\Users\\hj\\IdeaProjects\\Demo\\class"), parent);
System.out.println("parent 类加载器: " + parent);
System.out.println("classLoader1 类加载器: " + classLoader1);
System.out.println("classLoader2 类加载器: " + classLoader2);

Class<?> commonClazz = classLoader1.loadClass("Common");
System.out.println("commonClazz 类加载器:" + commonClazz.getClassLoader());
Object commonObj = commonClazz.newInstance();
```

```
// 将 classLoader1 或者 classLoader2 设置为线程的上下文类加载器
Thread.currentThread().setContextClassLoader(classLoader1);
Method commonTestMethod = commonClazz.getDeclaredMethod("test");
commonTestMethod.invoke(commonObj);
```

上述代码运行结果如下。

```
parent 类加载器: org.com.msb.classloader.Main$MyClassLoader@6d6f6e28
classLoader1 类加载器: org.com.msb.classloader.Main$MyClassLoader@135fbaa4
classLoader2 类加载器: org.com.msb.classloader.Main$MyClassLoader@45ee12a7
commonClazz 类加载器: org.com.msb.classloader.Main$MyClassLoader@6d6f6e28
need A class , start contextClassLoader load
load success
```

此时就能够完美地解决这种父类加载器需要通过子类加载器加载类的需求了。当然，这样也就打破了双亲委派机制。

6.2 路由断言使用

基于前面的代码，下面进行路由断言的学习。

6.2.1 Path 路由断言

Path 路由断言是根据 Path 定义好的规则来判断访问的 URL 是否匹配。下面通过实战来学习，请看如下代码。

```yaml
spring:
  application:
    name: cloud-gateway
  cloud:
    gateway:
      routes:
        - id: path
          predicates:
            - Path=/gateway/**
```

```yaml
      uri: lb://passenger-api/
    discovery:
      locator:
        enabled: true
```

如果请求路径为/gateway/多级路径，则匹配 lb://passenger-api/gateway/多级路径。例如，将 http://localhost:9001/gateway/hello/1 转发到 lb://passenger-api/gateway/hello。

6.2.2 Query 路由断言

Query 路由断言接收两个参数：一个是参数名，另一个是参数值。参数值可以是正则表达式。

```yaml
server:
  port: 9001
spring:
  application:
    name: cloud-gateway
  cloud:
    gateway:
      routes:
        - id: query
          predicates:
            - Query=name,cpf
          uri: lb://passenger-api/
      discovery:
        locator:
          enabled: true
eureka:
  client:
    service-url:
      defaultZone: http://localhost:8761/eureka/
  instance:
    instance-id: ${eureka.instance.hostname}:${server.port}
    hostname: localhost
```

上述配置，当请求路径中包含参数 name=cpf 时，进行匹配，否则不匹配。

Query 也可以支持正则表达式。来看如下配置。

```yaml
server:
  port: 9001
spring:
  application:
    name: cloud-gateway
  cloud:
    gateway:
      routes:
        - id: query
          predicates:
            - Query=name,cpf.
          uri: lb://passenger-api/
      discovery:
        locator:
          enabled: true
eureka:
  client:
    service-url:
      defaultZone: http://localhost:8761/eureka/
  instance:
    instance-id: ${eureka.instance.hostname}:${server.port}
    hostname: localhost
```

上述配置是匹配请求中带参数 name=cpfx（最后一位 x 可以是任意字母）的请求的。

6.2.3　Method 路由断言

Method 路由断言接收一个参数，匹配 HTTP 请求的方法。

```yaml
server:
  port: 9001
spring:
  application:
    name: cloud-gateway
  cloud:
    gateway:
```

```yaml
    routes:
      - id: query
        predicates:
          - Method=GET
        uri: lb://passenger-api/
      discovery:
        locator:
          enabled: true
eureka:
  client:
    service-url:
      defaultZone: http://localhost:8761/eureka/
  instance:
    instance-id: ${eureka.instance.hostname}:${server.port}
    hostname: localhost
```

当请求是 GET 时，匹配成功；当请求是 POST 时，请求失败。

6.2.4　Header 路由断言

Header 路由断言接收两个参数，即请求头名称和正则表达式。

```yaml
server:
  port: 9001
spring:
  application:
    name: cloud-gateway
  cloud:
    gateway:
      routes:
        - id: query
          predicates:
            - Header=header-name,cpf.
          uri: lb://passenger-api/
      discovery:
        locator:
          enabled: true
```

```yaml
eureka:
  client:
    service-url:
      defaultZone: http://localhost:8761/eureka/
  instance:
    instance-id: ${eureka.instance.hostname}:${server.port}
    hostname: localhost
```

测试 Header 路由断言，结果如图 6-10 所示。

图 6-10　Header 路由断言

6.2.5　自定义路由断言

当 Gateway 默认提供的路由断言无法满足使用时，可以自定义路由断言。

自定义路由断言需要继承 AbstractRoutePredicateFactory 类，重写 apply 方法。在 apply 方法中可以通过 serverWebExchange.getRequest() 拿到请求信息。

下面通过两个例子来实践一下。

（1）请求参数中有指定的名称才能匹配。

```
package com.cpf.cloud.gateway;

import org.springframework.cloud.gateway.handler.predicate.
```

```java
AbstractRoutePredicateFactory;
import org.springframework.stereotype.Component;
import org.springframework.web.server.ServerWebExchange;

import java.util.Arrays;
import java.util.List;
import java.util.function.Predicate;

/**
 * @author 马士兵教育:chaopengfei
 * @date 2022/3/22
 */
@Component
public class MyCheckRoutePredicateFactory extends
AbstractRoutePredicateFactory<MyCheckRoutePredicateFactory.MyConfig> {
    public MyCheckRoutePredicateFactory() {
        super(MyConfig.class);
    }
    // 将配置文件中的值按返回集合的顺序，赋值给配置类
    @Override
    public List<String> shortcutFieldOrder() {
        return Arrays.asList(new String[] {"name"});
    }


    @Override
    public Predicate<ServerWebExchange> apply(MyConfig config) {
        return (ServerWebExchange serverWebExchange) -> {
            System.out.println("自定义路由断言");
            String name = serverWebExchange.getRequest().getQueryParams().getFirst("name");
            if (name.equals(config.getName())){
                return true;
            }else {
                return false;
            }
        };
```

```java
    }

    public static class MyConfig{
        private String name;

        public String getName() {
            return name;
        }

        public void setName(String name) {
            this.name = name;
        }
    }
}
```

上述代码中下面的方法最重要,它将配置文件中的参数赋值给配置类。

```java
// 将配置文件中的值按返回集合的顺序,赋值给配置类
@Override
public List<String> shortcutFieldOrder() {
    return Arrays.asList(new String[] {"name"});
}
```

修改 application.yml。

```yaml
server:
  port: 9001
spring:
  application:
    name: cloud-gateway
  cloud:
    gateway:
      routes:
        - id: query
          predicates:
            - MyCheck=cpf1
          uri: lb://passenger-api/
      discovery:
        locator:
          enabled: true
```

```yaml
eureka:
  client:
    service-url:
      defaultZone: http://localhost:8761/eureka/
  instance:
    instance-id: ${eureka.instance.hostname}:${server.port}
    hostname: localhost
```

在配置 spring.cloud.gateway.routes[0].predicates[0]:MyCheck=cpf1 中，MyCheck 为类 MyCheckRoutePredicateFactory 名字中 RoutePredicateFactory 前面的部分，cpf1 为 MyConfig 中 name 的值。当请求 http://localhost:9001/gateway/hello?name=cpf1 时，路由转发成功。当请求 http://localhost: 9001/gateway/hello?name=cpf2 时，路由转发失败。

（2）请求参数中有指定的年龄范围才能匹配。

下面这个过滤器是判断请求中的年龄参数是否在配置文件中设置的范围内，如果在，则匹配成功，如果不在，则匹配失败。

```java
package com.cpf.cloud.gateway;

import org.springframework.cloud.gateway.handler.predicate.AbstractRoutePredicateFactory;
import org.springframework.stereotype.Component;
import org.springframework.web.server.ServerWebExchange;

import java.util.Arrays;
import java.util.List;
import java.util.function.Predicate;

/**
 * @author 马士兵教育:chaopengfei
 * @date 2022/3/22
 */
@Component
public class AgeCheckRoutePredicateFactory extends AbstractRoutePredicateFactory<AgeCheckRoutePredicateFactory.MyConfig> {
    public AgeCheckRoutePredicateFactory() {
        super(MyConfig.class);
    }
```

```java
    @Override
    public List<String> shortcutFieldOrder() {
        return Arrays.asList(new String[] {"minAge","maxAge"});
    }


    @Override
    public Predicate<ServerWebExchange> apply(MyConfig config) {
        return (ServerWebExchange serverWebExchange) -> {
            // 获取请求参数age,判断是否满足在yml中配置的值[1, 9)
            String age = serverWebExchange.getRequest().getQueryParams().getFirst("age");

            if (!age.matches("[0-9]+")) {
                return false;
            }

            int iAge = Integer.parseInt(age);

            if (iAge >= config.minAge && iAge < config.maxAge) {
                return true;
            } else {
                return false;
            }

        };
    }

    public static class MyConfig{
        private int minAge;
        private int maxAge;

        public int getMinAge() {
            return minAge;
        }

        public void setMinAge(int minAge) {
            this.minAge = minAge;
```

```
        }

        public int getMaxAge() {
            return maxAge;
        }

        public void setMaxAge(int maxAge) {
            this.maxAge = maxAge;
        }
    }
}
```

修改 aplication.yml。

```yaml
server:
  port: 9001
spring:
  application:
    name: cloud-gateway
  cloud:
    gateway:
      routes:
        - id: query
          predicates:
            - AgeCheck=1,9
          uri: lb://passenger-api/
      discovery:
        locator:
          enabled: true
eureka:
  client:
    service-url:
      defaultZone: http://localhost:8761/eureka/
  instance:
    instance-id: ${eureka.instance.hostname}:${server.port}
    hostname: localhost
```

请求 http://localhost:9001/gateway/hello?age=4 路由成功，因为 4 属于 1~9。

请求 http://localhost:9001/gateway/hello?age=40 路由失败，因为 40 不属于 1~9。

6.3 过滤器的使用

通过网关的请求，可以在网关给请求中添加一些信息，或者在请求的响应中添加一些信息。这个功能是通过 GatewayFilter 提供的，如果要实现更复杂的业务场景，也支持自定义。下面学习几种过滤器的使用。

6.3.1 添加请求头过滤器

在网关中修改 application.yml 文件，添加下面的配置。

```yaml
server:
  port: 9001
spring:
  application:
    name: cloud-gateway
  cloud:
    gateway:
      routes:
        - id: add_request_header
          predicates:
            - Path=/gateway/**
          uri: lb://passenger-api/
          filters:
            - AddRequestHeader=x-header-arg,add-value
      discovery:
        locator:
          enabled: true
eureka:
  client:
    service-url:
      defaultZone: http://localhost:8761/eureka/
  instance:
    instance-id: ${eureka.instance.hostname}:${server.port}
    hostname: localhost
```

来看下面这段配置。

```yaml
spring:
  cloud:
    gateway:
      routes:
        - id: add_request_header
          predicates:
            - Path=/gateway/**
          uri: lb://passenger-api/
          filters:
            - AddRequestHeader=x-header-arg,add-value
```

上述配置是对满足 Path 匹配条件的请求，给请求的 header 中添加一个参数 x-header-arg，参数值为 add-value。目的是在路由后的服务中做验证。

在 passenger-api 中，添加如下代码来获取上一步的 header 参数 x-header-arg。

```java
@GetMapping("/add-request")
public String addRequest(HttpServletRequest request){
    String header = request.getHeader("x-header-arg");
    return header;
}
```

经过测试，在返回结果中可以看到参数值 add-value。

6.3.2 移除请求头过滤器

做一个基本的路由断言，让匹配到的路径请求到后端的服务。

```yaml
server:
  port: 9001
spring:
  application:
    name: cloud-gateway
  cloud:
    gateway:
      routes:
        - id: add_request_header
          predicates:
```

```
        - Path=/gateway/**
          uri: lb://passenger-api/
      discovery:
        locator:
          enabled: true
eureka:
  client:
    service-url:
      defaultZone: http://localhost:8761/eureka/
  instance:
    instance-id: ${eureka.instance.hostname}:${server.port}
    hostname: localhost
```

在 Postman 中添加请求头 x-header-arg，值为 add-value。用 Postman 请求进行测试，结果如图 6-11 所示。

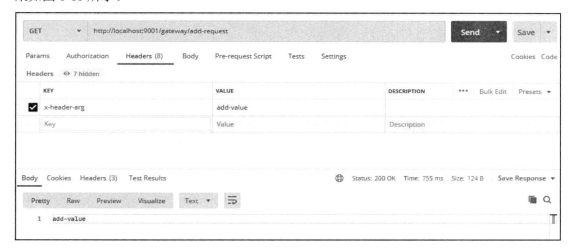

图 6-11　请求头设置成功

对网关进行如下改造。

```
server:
  port: 9001
spring:
  application:
    name: cloud-gateway
  cloud:
```

```yaml
    gateway:
      routes:
        - id: add_request_header
          predicates:
            - Path=/gateway/**
          uri: lb://passenger-api/
          filters:
            - RemoveRequestHeader=x-header-arg
      discovery:
        locator:
          enabled: true
eureka:
  client:
    service-url:
      defaultZone: http://localhost:8761/eureka/
  instance:
    instance-id: ${eureka.instance.hostname}:${server.port}
    hostname: localhost
```

上述代码中，在 filters 下增加 - RemoveRequestHeader=x-header-arg 配置，就会将请求中的 header 参数 x-header-arg 移除。用刚才的 Postman 测试发现，结果中的参数没有了，如图 6-12 所示。

图 6-12　请求头中的参数被移除

6.3.3 状态码设置

有的时候需要设置请求响应的状态码,可以通过下面的配置实现。

```yaml
server:
  port: 9001
spring:
  application:
    name: cloud-gateway
  cloud:
    gateway:
      routes:
        - id: add_request_header
          predicates:
            - Path=/gateway/**
          uri: lb://passenger-api/
          filters:
            - SetStatus=401
      discovery:
        locator:
          enabled: true
eureka:
  client:
    service-url:
      defaultZone: http://localhost:8761/eureka/
  instance:
    instance-id: ${eureka.instance.hostname}:${server.port}
    hostname: localhost
```

通过以上配置,在响应中的状态码就变成了 401。

6.3.4 重定向设置

在工作中,有时候需要将请求转发到其他的服务上,这时就需要用到重定向设置了。修改如下配置文件。

```yaml
server:
```

```yaml
    port: 9001
spring:
  application:
    name: cloud-gateway
  cloud:
    gateway:
      routes:
        - id: add_request_header
          predicates:
            - Path=/gateway/**
          uri: lb://passenger-api/
          filters:
            - RedirectTo=302,http://www.baidu.com
      discovery:
        locator:
          enabled: true
eureka:
  client:
    service-url:
      defaultZone: http://localhost:8761/eureka/
  instance:
    instance-id: ${eureka.instance.hostname}:${server.port}
    hostname: localhost
```

通过以上配置，就可以将请求转发到百度网站上去。

6.3.5 过滤器源码

过滤器其实和路由断言类似，查看它的源码实现，其实就可以自己写一个过滤器。例如，以上面用到的 AddRequestHeader 为例。打开 AddRequestHeaderGatewayFilterFactory，代码如下所示。

```java
public class AddRequestHeaderGatewayFilterFactory extends
AbstractNameValueGatewayFilterFactory {

    @Override
    public GatewayFilter apply(NameValueConfig config) {
```

```java
        return new GatewayFilter() {
            @Override
            public Mono<Void> filter(ServerWebExchange exchange, GatewayFilterChain chain) {
                String value = ServerWebExchangeUtils.expand(exchange, config.getValue());
                ServerHttpRequest request = exchange.getRequest().mutate().header(config.getName(), value).build();

                return chain.filter(exchange.mutate().request(request).build());
            }

            @Override
            public String toString() {
                return filterToStringCreator(AddRequestHeaderGatewayFilterFactory.this)
                        .append(config.getName(), config.getValue()).toString();
            }
        };
    }
```

以上代码中，关键的是 ServerHttpRequest request = exchange.getRequest().mutate().header(config.getName(), value).build();。从 yml 中获取参数，然后要通过这行代码将参数添加到 header 中。

6.4 全局过滤器

网关需要做一些全局性的限制，如限流、IP 黑名单等功能。因为这些是系统中通用的规则。这需要在网关中有一个统一的过滤器来完成。下面看一下全局过滤器的使用。

```
package com.cpf.cloud.gateway.filter;
```

```java
import org.springframework.cloud.gateway.filter.GatewayFilterChain;
import org.springframework.cloud.gateway.filter.GlobalFilter;
import org.springframework.context.annotation.Bean;
import org.springframework.context.annotation.Configuration;
import org.springframework.core.Ordered;
import org.springframework.core.annotation.Order;
import org.springframework.web.server.ServerWebExchange;
import reactor.core.publisher.Mono;

/**
 * @author 马士兵教育:chaopengfei
 * @date 2022/3/25
 */
@Configuration
public class GlobalCustomFilter{

    @Bean
    @Order(-1)
    public GlobalFilter zero(){
        return (exchange,chain)->{
            return chain.filter(exchange)
                .then(Mono.fromRunnable(()-> { System.out.println("-1 号过滤器");}));
        };
    }

    @Bean
    @Order(0)
    public GlobalFilter second(){
        return (exchange,chain)->{
            return chain.filter(exchange).then(Mono.fromRunnable(()-> System.out.println("0 号过滤器")));
        };
    }

    @Bean
    @Order(1)
```

```
public GlobalFilter first(){
    return (exchange,chain)->{
        return chain.filter(exchange).then(Mono.fromRunnable(()->
System.out.println("1 号过滤器")));
    };
}
```

上述代码中，从@Order(数字)来看，数字越小，越先执行。可以利用全局过滤器来实现一些系统的通用功能。

大厂面试

面试官：能否简单说一下你在工作中如何使用 Gateway，以及它的原理是什么？

回答分析：从两个方面来回答。

（1）回答自己在工作中如何使用 Gateway，体现自己丰富的实战经验。用网关进行统一鉴权服务、安全控制、服务限流、负载均衡、服务监控等。

（2）回答 Gateway 的主要原理。使用 RouteLocatorBuilder 的 bean 去创建路由，创建路由 RouteLocatorBuilder 可以让你添加各种 predicates 和 filters。predicates 是断言的意思，顾名思义，就是根据具体的请求的规则，由具体的 route 去处理；filters 是指各种过滤器，用来对请求做各种判断和修改。

6.5 小　　结

本章学习了 Gateway 网关的基本使用，包括没有注册中心和结合注册中心的使用。学习了路由断言的用法，包括 Path、Query、Method、Header 等路由断言，主要是判断路由的规则。学习了过滤器的使用，包括通过给请求中添加 header 信息、删除响应中的 header 信息、重定向等，主要是为了对请求进行改造。通过本章的学习，能让读者达到在生产环境中使用网关的能力。

第 7 章 链路追踪

在微服务架构中,微服务之间的调用链路特别多,尤其大项目有成百上千个服务互相调用,如果调用链路中的某一个环节出了问题,那么整个调用链都会出问题,排查问题是一件相当困难的事情。这时就需要追踪每个调用链的情况,Spring Cloud 提供的 Sleuth 可以完美地解决这个问题。本章就来学习如何通过 Sleuth 进行链路追踪,以及如何通过可视化界面来展示链路追踪的信息。

7.1 链路追踪的设计思路

像前面一样,从设计者的角度来分析。如果让开发人员来实现这个需求,需要怎么做?业务需求是需要记录服务之间调用是否成功,以及耗时时长。

就类似于 A 给 B 打电话,让 B 给 C 打电话,来获得某个信息,最终 A 没有收到响应。此时需要确定这中间是 A 给 B 打电话出了问题,还是 B 给 C 打电话出了问题。应该如何分析呢?

办法简单,记录下 A 和 B 的行为轨迹。例如,A 什么时间接收到命令,然后什么时间给 B 打了电话,B 什么时间收到 A 的电话,B 什么时间给 C 打电话,以及他们什么时间收到对方的回复。

顺着这条思路下来,是不是就能分析出在每次接口请求时,需要记录哪些数据了呢?

- ☑ 服务名称:执行业务的服务名称。
- ☑ 全局请求 ID:在所有服务共同完成一项任务时,这项任务在整个项目中的唯一 ID。
- ☑ 工作单元:服务发起一次远程调用,就是一个工作单元。
- ☑ 请求时间:每次工作单元执行的时间。

7.2 链路追踪的使用

1. 服务准备

准备一个注册中心 cloud-eureka,这还是前面章节中的内容,在此不再赘述。然后以两个 Web 服务调用为例,即 passenger-api 调用 pay-service。

pay-service 提供接口/provider/test,输出 pay restful provider。

```
@RestController
@RequestMapping("/provider")
public class ProviderController {

    @GetMapping("/test")
    public String test() {
        System.out.println(port);
        return "pay restful provider";
    }
}
```

passenger-api 调用 pay-service 上面提供的接口,最终输出 pay restful provider。

```
@RestController
@RequestMapping("/restful")
public class RestTemplateController {

    @Autowired
    private RestTemplate restTemplate;

    @GetMapping("/test")
    public String test() {
        // 此处注意,一直写的 IP 和端口,现在是服务名(其实本质是虚拟主机名):pay-service
        String url = "http://pay-service/provider/test";
        return restTemplate.getForObject(url, String.class);
    }
}
```

2. 引入 Sleuth 的 jar 包

在 passenger-api 和 pay-service 中分别引入 Sleuth 的 jar 包。

```xml
<dependency>
    <groupId>org.springframework.cloud</groupId>
    <artifactId>spring-cloud-starter-sleuth</artifactId>
</dependency>
```

在 application.yml 中分别修改两个项目的日志级别。

```yaml
logging:
  level:
    org.springframework.web.servlet.DispatcherServlet: debug
```

3. 测试请求并观察日志

调用 passenger-api 的请求 http://localhost:8090/restful/test。请求成功后，查看两个项目的日志如下。

```
// passenger-api 日志：
2021-03-25 13:04:53.051 DEBUG [passenger-api,a47ddee37c6c4ad4,
a47ddee37c6c4ad4] 23344 --- [nio-8090-exec-3] o.s.web.servlet.
DispatcherServlet         : GET "/restful/test", parameters={}
2021-03-25 13:04:53.060 DEBUG [passenger-api,a47ddee37c6c4ad4,
a47ddee37c6c4ad4] 23344 --- [nio-8090-exec-3] o.s.web.servlet.
DispatcherServlet         : Completed 200 OK
```

观察到[passenger-api,a47ddee37c6c4ad4,a47ddee37c6c4ad4]，第一个值 passenger-api 就是当前的服务名称 passenger-api；第二个值 a47ddee37c6c4ad4 是全局 traceID，代表这一次请求的全局 ID，即唯一标识；第三个值 a47ddee37c6c4ad4 是 spanID，表示当前的服务基本单元。

```
// pay-service 日志：
2021-03-25 13:04:53.055 DEBUG [pay-service,a47ddee37c6c4ad4,
ee9b868a86225cd3] 18632 --- [nio-8092-exec-3] o.s.web.servlet.
DispatcherServlet         : GET "/provider/test", parameters={}
2021-03-25 13:04:53.058 DEBUG [pay-service,a47ddee37c6c4ad4,
ee9b868a86225cd3] 18632 --- [nio-8092-exec-3] o.s.web.servlet.
DispatcherServlet         : Completed 200 OK
```

观察到[pay-service,a47ddee37c6c4ad4,ee9b868a86225cd3]，第一个值 pay-service 就是当前的服务名称 pay-service；第二个值 a47ddee37c6c4ad4 是全局 traceID，代表这一次请求的全局 ID，即唯一标识；第三个值 ee9b868a86225cd3 是 spanID，表示当前的服务基本单元。

在这一次调用链路中，a47ddee37c6c4ad4 是全局 traceID，在 passenger-api 和 pay-service 中都是一样的。这样就将这一次的请求给串联起来了。而 passenger-api 的 spanID 是 a47ddee37c6c4ad4，pay-service 的 spanID 是 ee9b868a86225cd3。

其实这里的日志内容，也就是 7.1 节中分析的每次接口请求时需要记录的数据。

- 第一个值：服务名称，就是当前执行业务的服务名称。
- 第二个值：这是 Sleuth 生成的一个全局追踪 ID，贯穿整个调用链路，称为 traceID。一个链路中只有一个 traceID。
- 第三个值：是一个工作单元的标识，如 passenger-api 调用 pay-server 为一个工作单元。

其中，traceID 和 spanID 是业务的关键信息，在一次请求的整个调用链路中，都是通过 traceID 将所有的 spanID 串联起来的。

7.3 追踪原理分析

根据 7.2 节中讲到的链路追踪的基本使用，来分析它的实现原理。其核心数据为以下两部分。

- 全局 ID：当一次请求发生时，需要 Sleuth 框架为这次请求分配一个全局 traceID，当服务间互相调用时，在发出去的请求中将 traceID 作为参数带上，直到请求返回。这样就能将所有的请求串联起来。
- 统计部分单元情况。当请求到达各个服务时，也要通过每个服务的特殊标记，如 spanID 用来记录请求的状态，以及记录开始、执行、结束的时间。通过每个 spanID 的执行时间差，就可以统计出延时时长。

在前面的例子中，passenger-api 调用 pay-service 是通过 RestTemplate 调用完成的。Sleuth 组件会在请求调用时，对请求进行处理。Sleuth 会改造调用的请求头，在 header 中添加参数，主要有以下 3 个参数。

- X-B3-TraceId：请求链路中的全局标识，也就是 traceID。
- X-B3-SpanId：工作单元的唯一标识，也就是 spanID。
- X-B3-ParentSpanId：当前工作单元的上一个工作单元。如果是 Root Span，则该值为空。

下面通过改造代码来验证一下。在 pay-service 中从请求头中将上面 3 个参数打印出来。

```
@GetMapping("/test")
public String test(HttpServletRequest request) {
    log.info("X-B3-TraceId:"+request.getHeader("X-B3-TraceId"));
    log.info("X-B3-SpanId:"+request.getHeader("X-B3-SpanId"));
    log.info("X-B3-ParentSpanId:"+request.getHeader("X-B3-ParentSpanId"));

    System.out.println(port);
    return "pay restful provider";
}
```

执行一次请求，然后观察日志。

```
2021-03-25 13:36:34.771 DEBUG [pay-service,aadd642f6874f63c,a5b30c4896bbce06] 10588 --- [nio-8092-exec-1] o.s.web.servlet.DispatcherServlet        : GET "/provider/test", parameters={}
2021-03-25 13:36:34.784  INFO [pay-service,aadd642f6874f63c,a5b30c4896bbce06] 10588 --- [nio-8092-exec-1] c.cpf.pay.controller.ProviderController  : X-B3-TraceId:aadd642f6874f63c
2021-03-25 13:36:34.784  INFO [pay-service,aadd642f6874f63c,a5b30c4896bbce06] 10588 --- [nio-8092-exec-1] c.cpf.pay.controller.ProviderController  : X-B3-SpanId:4b4076c62d6a3b22
2021-03-25 13:36:34.784  INFO [pay-service,aadd642f6874f63c,a5b30c4896bbce06] 10588 --- [nio-8092-exec-1] c.cpf.pay.controller.ProviderController  : X-B3-ParentSpanId:aadd642f6874f63c
```

发现以上日志结果和分析的一样。

7.4 可视化链路追踪

通过日志可以分析出请求链路的调用情况，而在实际开发中，如果使用这样的处理方式，还是不方便，因为要一直查看日志，如果日志特别多，也不现实，反而增加了工作量，效果还不一定好。如果能可视化的观察链路调用的结果就好了，这时可以选用 Zipkin。

Zipkin 是一个开源项目，它主要提供了收集数据和查询数据两大接口服务，有了 Zipkin，就能直观地对调用链路结果进行查看，并且可以非常方便地看到服务直接的调用关系，以及调用的耗时情况。下面我们来学习使用这个工具。

1. 下载安装 Zipkin

打开 Zipkin 的官网 https://zipkin.io/pages/quickstart.html，如图 7-1 所示。

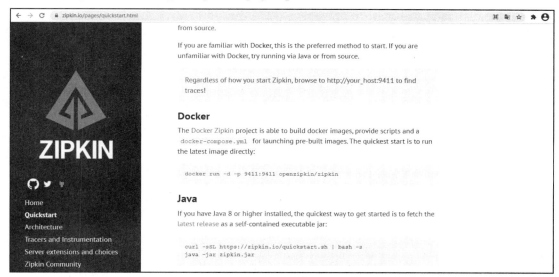

图 7-1　Zipkin 官网

用 Java 程序的启动方式来运行。执行下面的命令来下载 zipkin.jar 包。

```
curl -sSL https://zipkin.io/quickstart.sh | bash -s
```

命令执行完之后，会得到一个 zipkin.jar 包。

2. 启动 Zipkin

执行如下命令启动 Zipkin。

```
java -jar zipkin.jar
```

启动成功界面如图 7-2 所示。

3. 配置服务

前面讲过将追踪信息输出到日志中，现在已经有了 Zipkin 可视化界面，需要将链路追踪信息发送给 Zipkin，让 Zipkin 收集起来，并进行可视化展示。

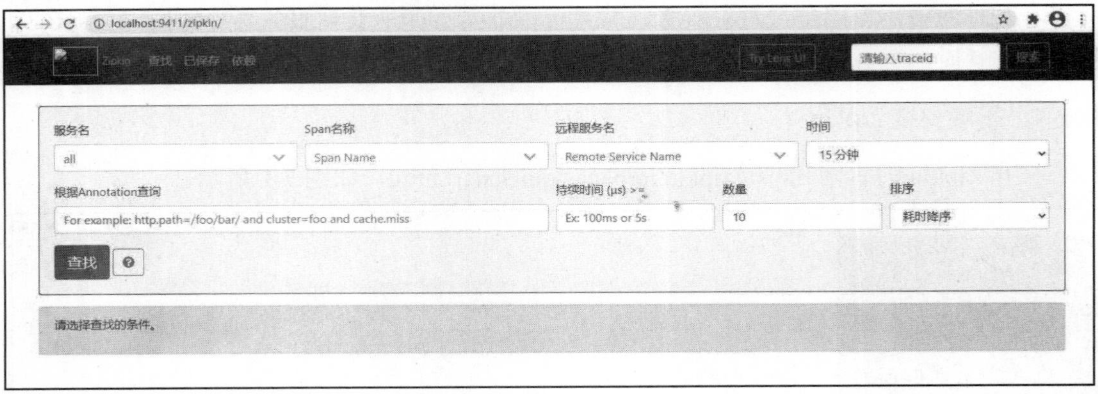

图 7-2 Zipkin 启动成功

在每个服务的 pom 中添加 Zipkin 依赖。

```
<!-- https://mvnrepository.com/artifact/org.springframework.cloud/spring-cloud-starter-zipkin -->
<dependency>
    <groupId>org.springframework.cloud</groupId>
    <artifactId>spring-cloud-starter-zipkin</artifactId>
    <version>2.1.0.RELEASE</version>
</dependency>
```

修改每个服务的 application.yml。

```
server:
  port: 8090
spring:
  application:
    name: passenger-api

  zipkin:
    base-url: http://localhost:9411
    discovery-client-enabled: false
  sleuth:
    sampler:
      probability: 1
```

```
eureka:
  client:
    serviceUrl:
      defaultZone: http://localhost:8761/eureka/
  instance:
    hostname: localhost
logging:
  level:
    org.springframework: debug
```

上面的 application.yml 文件中的两个配置说明如下。

☑ spring.zipkin.base-url：配置了 Zipkin 的地址。

☑ spring.sleuth.sampler.probability：配置采样率，1 表示将日志数据 100%采样。

调用一次请求 http://localhost:8090/restful/test。观察 Zipkin 界面，如图 7-3 所示。

图 7-3　Zipkin 链路追踪数据界面

查看调用详情，如图 7-4 所示。

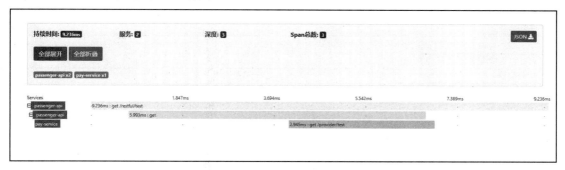

图 7-4　Zipkin 调用详情

从图 7-4 可以看到，请求进来先被 passenger-api 接收，然后 passenger 调用了 pay-service 的/provider/test 方法。

在实际生产中，如果采样比例是 1 的话，也就是 100%采集链路追踪数据，这样对服务的性能会造成影响。一般情况下，在项目的前期会将采样比例设置成 1，用来观察系统的链路情况。等系统稳定后，一般将采样比例调整成 0.1，也就是 10%的采样比例。

下面通过演示一个异常信息来观察 Zipkin 的表现。

在 pay-service 的接口中手动写一个异常。

```
@GetMapping("/test")
public String test(HttpServletRequest request) {
    log.info("X-B3-TraceId:"+request.getHeader("X-B3-TraceId"));
    log.info("X-B3-SpanId:"+request.getHeader("X-B3-SpanId"));
    log.info("X-B3-ParentSpanId:"+request.getHeader("X-B3-ParentSpanId"));
    // 异常
    int i = 1/0;
    return "pay restful provider";
}
```

调用接口 http://localhost:8090/restful/test。观察 Zipkin 异常，如图 7-5 所示。

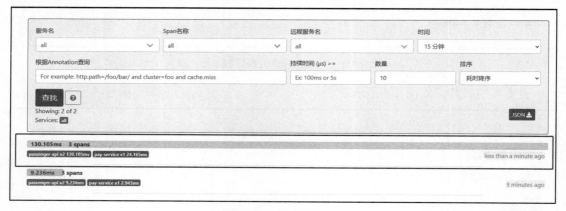

图 7-5　Zipkin 异常

可以看到图 7-5 中的界面有一个系统异常的提示。

7.5 消息队列收集链路追踪

用 HTTP 进行采样收集，对性能会有影响。如果 Zipkin 服务重启或停止了，就会丢失采样数据。为了解决这个问题，需要集成 RabbitMQ 来进行数据采集，利用消息队列提高系统的性能，保证数据不丢失。

（1）在每个服务中增加对 RabbitMQ 的依赖。

```xml
<dependency>
    <groupId>org.springframework.boot</groupId>
    <artifactId>spring-boot-starter-amqp</artifactId>
</dependency>
```

（2）修改 application.yml。

```yaml
server:
  port: 8090
spring:
  application:
    name: passenger-api

  zipkin:
    sender:
      type: rabbit
  rabbitmq:
    addresses: localhost:5672
    username: guest
    password: guest
  sleuth:
    sampler:
      probability: 1

eureka:
  client:
    serviceUrl:
      defaultZone: http://localhost:8761/eureka/
```

```
    instance:
      hostname: localhost
logging:
  level:
    org.springframework: debug
```

在 application.yml 中修改了以下两个配置。
- ☑ spring.zipkin.sender.type 设置成 rabbit。
- ☑ 配置 RabbitMQ 的相关信息。

修改完成后，重新启动两个服务。

（3）修改 Zipkin 的启动方式。

```
java -DRABBIT_ADDRESSES=localhost:5672 -DRABBIT_USER=guest -DRABBITIT_PASSWORD=guest -jar zipkin.jar
```

在启动 Zipkin 时，指定对应的消息队列。

7.6 小　　结

本章学习了如何在 Spring Cloud 中进行链路追踪。介绍了 Sleuth 的使用方法和原理，通过日志的打印，知道了采集的数据信息，以及可视化观测组件 Zipkin。关于 Zipkin 介绍了两种收集链路信息的方式，一种是 HTTP，另外一种是消息队列。

第 8 章 服务监控

在实际生产中需要对服务进行健康检查,以便及时发现有问题的服务,并快速进行处理,避免对业务造成影响。当服务较少时,可以在服务上写脚本定时执行任务,检查 CPU、内存和磁盘的情况。当服务特别多时,需要一个组件能可视化地、统一地对所有的服务进行检测。这就是 Spring Boot Admin 组件,它是非常好用的监控和管理软件,能够将服务的 actuator 模块中的端点信息进行收集,并提供可视化的展示,还可以提供报警功能。

8.1 Spring Boot Admin 的使用

1. 服务准备

准备两个 Web 项目,项目如下。
- ☑ passenger-api:乘客接口服务。
- ☑ cloud-eureka:注册中心。

对 passenger-api 服务的健康状况进行监控,需要启动 cloud-eureka 和 passenger-api,如图 8-1 所示。

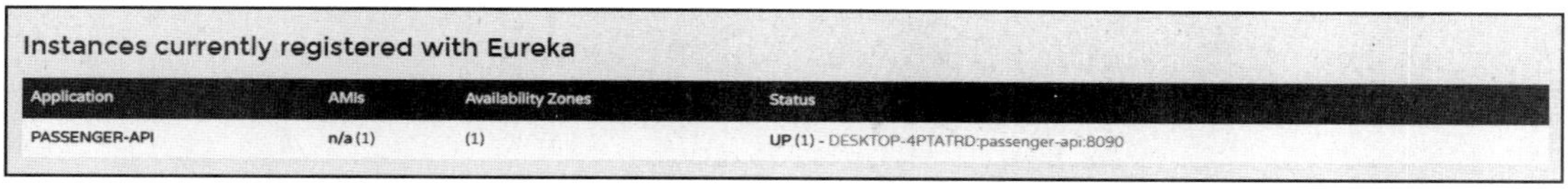

Instances currently registered with Eureka			
Application	AMIs	Availability Zones	Status
PASSENGER-API	n/a (1)	(1)	UP (1) - DESKTOP-4PTATRD:passenger-api:8090

图 8-1 cloud-eureka 和 passenger-api 启动

2. 创建 Admin 服务

新建一个 Web 项目 cloud-admin-server。在 pom 中引入如下依赖。

```
<dependencies>
    <dependency>
```

```xml
        <groupId>org.springframework.boot</groupId>
        <artifactId>spring-boot-starter-web</artifactId>
    </dependency>

    <dependency>
        <groupId>de.codecentric</groupId>
        <artifactId>spring-boot-admin-starter-server</artifactId>
    </dependency>

    <dependency>
        <groupId>org.springframework.boot</groupId>
        <artifactId>spring-boot-starter-test</artifactId>
        <scope>test</scope>
    </dependency>
</dependencies>
```

上述代码中，de.codecentric:spring-boot-admin-starter-server 是 admin 的依赖。

修改启动类，添加注解@EnableAdminServer。

```java
@SpringBootApplication
@EnableAdminServer
public class CloudAdminServerApplication {

    public static void main(String[] args) {
        SpringApplication.run(CloudAdminServerApplication.class, args);
    }

}
```

修改 application.yml。

```yaml
server:
  port: 9091
spring:
  application:
    name: cloud-admin-server
```

启动 cloud-admin-server，访问 http://localhost:9091/，Admin 启动界面如图 8-2 所示。

图 8-2 Admin 启动界面

3. 将 cloud-admin-server 注册到注册中心

在 pom 中添加 eureka-client 的依赖。

```
<dependency>
    <groupId>org.springframework.cloud</groupId>
    <artifactId>spring-cloud-starter-netflix-eureka-client</artifactId>
</dependency>
```

最终的 pom 依赖如下。

```
<?xml version="1.0" encoding="UTF-8"?>
<project xmlns="http://maven.apache.org/POM/4.0.0" xmlns:xsi="http://
www.w3.org/2001/XMLSchema-instance" xsi:schemaLocation="http://maven.
apache.org/POM/4.0.0 https://maven.apache.org/xsd/maven-4.0.0.xsd">
    <modelVersion>4.0.0</modelVersion>
    <parent>
        <groupId>org.springframework.boot</groupId>
        <artifactId>spring-boot-starter-parent</artifactId>
        <version>2.4.3</version>
        <relativePath/> <!-- lookup parent from repository -->
    </parent>
    <groupId>com.cpf</groupId>
    <artifactId>cloud-admin-server</artifactId>
    <version>0.0.1-SNAPSHOT</version>
    <name>cloud-admin-server</name>
    <description>Demo project for Spring Boot</description>
    <properties>
        <java.version>1.8</java.version>
        <spring-boot-admin.version>2.3.1</spring-boot-admin.version>
```

```xml
        <spring-cloud.version>2020.0.1</spring-cloud.version>
    </properties>
    <dependencies>
        <dependency>
            <groupId>org.springframework.boot</groupId>
            <artifactId>spring-boot-starter-web</artifactId>
        </dependency>

        <dependency>
            <groupId>de.codecentric</groupId>
            <artifactId>spring-boot-admin-starter-server</artifactId>
        </dependency>

        <dependency>
            <groupId>org.springframework.boot</groupId>
            <artifactId>spring-boot-starter-test</artifactId>
            <scope>test</scope>
        </dependency>
    </dependencies>

    <dependencyManagement>
        <dependencies>
            <dependency>
                <groupId>de.codecentric</groupId>
                <artifactId>spring-boot-admin-dependencies</artifactId>
                <version>${spring-boot-admin.version}</version>
                <type>pom</type>
                <scope>import</scope>
            </dependency>

            <dependency>
                <groupId>org.springframework.cloud</groupId>
                <artifactId>spring-cloud-dependencies</artifactId>
                <version>${spring-cloud.version}</version>
                <type>pom</type>
                <scope>import</scope>
```

```xml
            </dependency>
        </dependencies>
    </dependencyManagement>

    <build>
        <plugins>
            <plugin>
                <groupId>org.springframework.boot</groupId>
                <artifactId>spring-boot-maven-plugin</artifactId>
            </plugin>
        </plugins>
    </build>
</project>
```

修改 application.yml 文件。

```yaml
server:
  port: 9091
spring:
  application:
    name: cloud-admin-server

eureka:
  client:
    serviceUrl:
      defaultZone: http://localhost:8761/eureka/
```

重新启动 cloud-admin-server，观察注册中心管理界面，如图 8-3 所示。

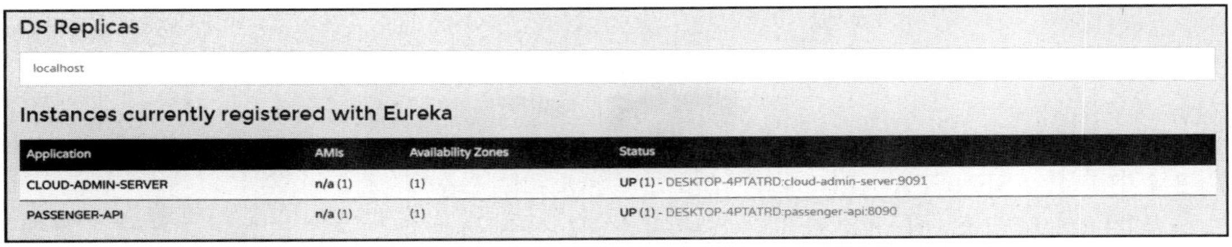

图 8-3 注册中心界面

4. 添加 actuator

在 cloud-admin-server 和 passenger-api 两个服务的 pom 中添加以下依赖。

```xml
<dependency>
    <groupId>org.springframework.boot</groupId>
    <artifactId>spring-boot-starter-actuator</artifactId>
    <version>2.3.0.RELEASE</version>
</dependency>
```

修改两个服务的 application.yml，暴露端点。

```yaml
management:
  endpoints:
    web:
      exposure:
        include: "*"
```

重新启动 cloud-admin-server 和 passenger-api 两个服务，Admin 监控界面如图 8-4 所示。

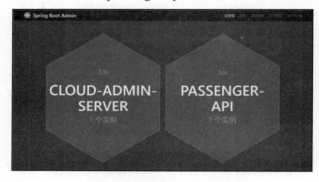

图 8-4　Admin 监控界面

8.2　监控内容介绍

8.1 节搭建了 Admin 监控服务，本节来了解 Admin 都能查看哪些监控的内容。单击图 8-4 中的 PASSENGER-API，监控详情如图 8-5 和图 8-6 所示。

从图 8-5 和图 8-6 中可以看出，监控的内容包括服务信息、健康状态、元数据、进程、线程、垃圾回收情况、堆内存和非堆内存的使用。

图 8-5　监控详情一

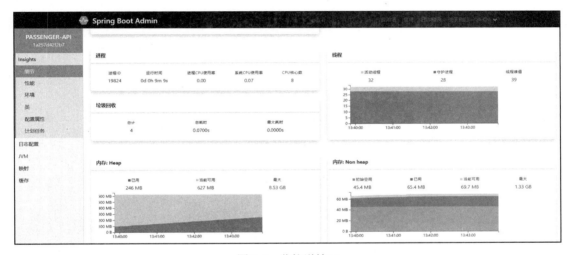

图 8-6　监控详情二

8.3　认 证 保 护

在访问健康检查服务时，最好做一层认证，最好不要让用户输入 URL 直接就可以访问，必须要输入用户名和密码，这样可以保证数据安全。下面对 cloud-admin-server 添加认证。

在 pom 中添加如下依赖。

```
<!-- 安全认证 -->
<dependency>
```

```
        <groupId>org.springframework.boot</groupId>
        <artifactId>spring-boot-starter-security</artifactId>
</dependency>
```

修改 application.yml。

```yaml
server:
  port: 9091
spring:
  application:
    name: cloud-admin-server
  security:
    user:
      name: root
      password: root

eureka:
  client:
    serviceUrl:
      defaultZone: http://localhost:8761/eureka/
  instance:
    hostname: localhost

management:
  endpoints:
    web:
      exposure:
        include: "*"
```

重新启动 cloud-admin-server。再访问 http://localhost:9091/时，就需要输入用户名和密码，才能进行访问了。

8.4 服务监听邮件通知

通过前面的学习，可以监控服务的健康状态。如果服务发生故障，还需要即时收到通知。在生产中的很多时候，我们希望能够自动监控，即通过邮件报警，或者钉钉通知的方式，即时获取服务下线的信息。

要实现服务下线后进行邮件通知，可以在 cloud-admin-server 的 pom 中添加如下依赖。

```xml
<dependency>
    <groupId>org.springframework.boot</groupId>
    <artifactId>spring-boot-starter-mail</artifactId>
</dependency>
```

修改 application.yml。

```yaml
spring:
  application:
    name: cloud-admin-server
  security:
    user:
      name: root
      password: root
  mail:
    host: smtp.qq.com
    username: chaopengfei100@qq.com
    #授权码
    password: qq邮箱授权码
    properties:
      mail:
        smtp:
          auth: true
          starttls:
            enable: true
            required: true
boot:
  admin:
    notify:
      mail:
        #发件人
        from: chaopengfei100@qq.com
        #收件人
        to: chaopengfei100@qq.com
        #配置是否启用邮件通知，false是不启用
        enabled: true
```

重启 cloud-admin-server。当服务上线时，就可以在邮箱中看到如图 8-7 所示的报警邮件。

```
CLOUD-ADMIN-SERVER (fd5e7f1ffc79) is UP

Instance fd5e7f1ffc79 changed status from OFFLINE to UP

Status Details

Registration

Service Url      http://localhost:9091
Health Url       http://localhost:9091/actuator/health
Management Url   http://localhost:9091/actuator
```

图 8-7　报警邮件

8.5　服务监听钉钉通知

设置钉钉智能机器人用于服务监听，先打开钉钉智能群助手，如图 8-8 所示。

图 8-8　智能群助手

单击"添加机器人"，如图 8-9 所示。

图 8-9 添加机器人

单击"自定义",添加机器人,如图 8-10 所示。

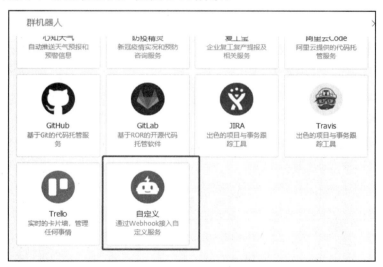

图 8-10 单击"自定义"

添加机器人信息之后单击"完成"按钮,如图 8-11 所示。

图 8-11 添加机器人信息

再次单击"完成"按钮，完成机器人的添加，如图 8-12 所示。

图 8-12 完成机器人添加

改造代码，在项目中添加如下依赖。

```
<dependencies>
    <dependency>
```

```xml
    <groupId>org.springframework.boot</groupId>
    <artifactId>spring-boot-starter-web</artifactId>
</dependency>

<dependency>
    <groupId>de.codecentric</groupId>
    <artifactId>spring-boot-admin-starter-server</artifactId>
</dependency>

<dependency>
    <groupId>org.springframework.boot</groupId>
    <artifactId>spring-boot-starter-test</artifactId>
    <scope>test</scope>
</dependency>

<dependency>
    <groupId>org.springframework.cloud</groupId>
    <artifactId>spring-cloud-starter-netflix-eureka-client</artifactId>
</dependency>

<dependency>
    <groupId>org.springframework.boot</groupId>
    <artifactId>spring-boot-starter-actuator</artifactId>
    <version>2.3.0.RELEASE</version>
</dependency>

<!-- 安全认证 -->
<dependency>
    <groupId>org.springframework.boot</groupId>
    <artifactId>spring-boot-starter-security</artifactId>
</dependency>

<dependency>
    <groupId>org.springframework.boot</groupId>
    <artifactId>spring-boot-starter-mail</artifactId>
</dependency>
```

```xml
<dependency>
    <groupId>com.aliyun</groupId>
    <artifactId>alibaba-dingtalk-service-sdk</artifactId>
    <version>1.0.1</version>
</dependency>
<dependency>
    <groupId>com.alibaba</groupId>
    <artifactId>fastjson</artifactId>
    <version>1.2.69</version>
</dependency>

<dependency>
    <groupId>org.projectlombok</groupId>
    <artifactId>lombok</artifactId>
    <version>1.18.12</version>
</dependency>

<dependency>
    <groupId>org.apache.httpcomponents</groupId>
    <artifactId>httpclient</artifactId>
    <version>4.5.13</version>
</dependency>

<dependency>
    <groupId>org.apache.httpcomponents</groupId>
    <artifactId>httpcore</artifactId>
    <version>4.4.13</version>
</dependency>

</dependencies>
```

编写钉钉发送工具类 DingtalkUtils。

```java
import com.alibaba.fastjson.JSON;
import lombok.extern.slf4j.Slf4j;
import org.apache.http.HttpResponse;
import org.apache.http.client.HttpClient;
import org.apache.http.client.methods.HttpPost;
```

```java
import org.apache.http.entity.StringEntity;
import org.apache.http.impl.client.HttpClients;
import org.apache.http.util.EntityUtils;
import org.springframework.http.HttpStatus;

import java.util.HashMap;

/**
 * 钉钉发送工具类
 */
@Slf4j
public class DingtalkUtils {

    public static void main(String[] args) {
        pushInfoToDingding("测试消息通知",
"30709c7b3f38101b853d1c50a211031d2ad358c1dca40e57d1c6f8435938f56c");
    }

    public static Boolean pushInfoToDingding(String textMsg, String dingURL) {

        HashMap<String, Object> resultMap = new HashMap<>(8);
        resultMap.put("msgtype", "text");

        HashMap<String, String> textItems = new HashMap<>(8);
        textItems.put("content", textMsg);
        resultMap.put("text", textItems);

        HashMap<String, Object> atItems = new HashMap<>(8);
        atItems.put("atMobiles", null);
        atItems.put("isAtAll", false);
        resultMap.put("at", atItems);


        dingURL = "https://oapi.dingtalk.com/robot/send?access_token=" + dingURL;
        try {
            HttpClient httpClient = HttpClients.createDefault();
```

```java
            StringEntity stringEntity = new StringEntity(JSON.toJSONString(resultMap), "utf-8");

            HttpPost httpPost = createConnectivity(dingURL);
            httpPost.setEntity(stringEntity);
            HttpResponse response = httpClient.execute(httpPost);
            if (response.getStatusLine().getStatusCode() == HttpStatus.OK.value()) {
                String result = EntityUtils.toString(response.getEntity(), "utf-8");
                System.out.println(result);
                log.info("执行结果: {}" , result);
            }
            return Boolean.TRUE;
        } catch (Exception e) {
            e.printStackTrace();
            return Boolean.FALSE;
        }
    }


    static HttpPost createConnectivity(String restUrl) {
        HttpPost post = new HttpPost(restUrl);
        post.setHeader("Content-Type", "application/json");
        post.setHeader("Accept", "application/json");
        post.setHeader("X-Stream", "true");
        return post;
    }
}
```

编写监听类 DingtalkNotifier。

```
package com.cpf.cloud.admin.server.ding;

import com.alibaba.fastjson.JSONObject;
import de.codecentric.boot.admin.server.domain.entities.Instance;
import de.codecentric.boot.admin.server.domain.entities.InstanceRepository;
```

```java
import de.codecentric.boot.admin.server.domain.events.InstanceEvent;
import de.codecentric.boot.admin.server.domain.events.InstanceStatusChangedEvent;
import de.codecentric.boot.admin.server.notify.AbstractStatusChangeNotifier;
import lombok.extern.slf4j.Slf4j;
import org.springframework.stereotype.Component;
import reactor.core.publisher.Mono;

import java.util.Arrays;

/**
 * 服务状态变更监听类
 */
@Slf4j
@Component
public class DingtalkNotifier extends AbstractStatusChangeNotifier {

    /**
     * 消息模板
     */
    private static final String template = "<<<%s>>> \n 【服务名】: %s(%s) \n 【状态】: %s(%s) \n 【服务ip】: %s \n 【详情】: %s";

    private String titleAlarm = "系统告警";

    private String titleNotice = "系统通知";

    private String[] ignoreChanges = new String[]{"UNKNOWN:UP", "DOWN:UP", "OFFLINE:UP"};

    public DingtalkNotifier(InstanceRepository repository) {
        super(repository);
    }

    @Override
    protected boolean shouldNotify(InstanceEvent event, Instance instance) {
```

```java
        if (!(event instanceof InstanceStatusChangedEvent)) {
            return false;
        } else {
            InstanceStatusChangedEvent statusChange = (InstanceStatusChangedEvent) event;
            String from = this.getLastStatus(event.getInstance());
            String to = statusChange.getStatusInfo().getStatus();
            return Arrays.binarySearch(this.ignoreChanges, from + ":" + to) < 0 && Arrays.binarySearch(this.ignoreChanges, "*:" + to) < 0 && Arrays.binarySearch(this.ignoreChanges, from + ":*") < 0;
        }
    }

    @Override
    protected Mono<Void> doNotify(InstanceEvent event, Instance instance) {

        return Mono.fromRunnable(() -> {

            if (event instanceof InstanceStatusChangedEvent) {
                log.info("Instance {} ({}) is {}", instance.getRegistration().getName(), event.getInstance(),((InstanceStatusChangedEvent) event).getStatusInfo().getStatus());
                String status = ((InstanceStatusChangedEvent) event).getStatusInfo().getStatus();
                String messageText = null;
                switch (status) {
                    // 健康检查没通过
                    case "DOWN":
                        log.info("发送 健康检查没通过 的通知！");
                        messageText = String.format(template, titleAlarm, instance.getRegistration().getName(), event.getInstance(), ((InstanceStatusChangedEvent) event).getStatusInfo().getStatus(), "健康检查没通过通知", instance.getRegistration().getServiceUrl(), JSONObject.toJSONString(instance.getStatusInfo().getDetails()));
                        // 先输出信息在控制台
                        DingtalkUtils.pushInfoToDingding(messageText, "30709c7b3f38101b853d1c50a211031d2ad358c1dca40e57d1c6f8435938f56c");
```

```
                break;
            // 服务离线
            case "OFFLINE":
                log.info("发送 服务离线 的通知！");
                messageText = String.format(template, titleAlarm,
instance.getRegistration().getName(), event.getInstance(),
((InstanceStatusChangedEvent) event).getStatusInfo().getStatus(), "服务离
线通知",instance.getRegistration().getServiceUrl(), JSONObject.
toJSONString(instance.getStatusInfo().getDetails()));
                // 先输出信息在控制台
                DingtalkUtils.pushInfoToDingding(messageText,
"30709c7b3f38101b853d1c50a211031d2ad358c1dca40e57d1c6f8435938f56c");
                break;
            // 服务上线
            case "UP":
                log.info("发送 服务上线 的通知！");
                messageText = String.format(template, titleNotice,
instance.getRegistration().getName(), event.getInstance(),
((InstanceStatusChangedEvent) event).getStatusInfo().getStatus(), "服务上
线通知",instance.getRegistration().getServiceUrl(), JSONObject.
toJSONString(instance.getStatusInfo().getDetails()));
                // 先输出信息在控制台
                System.out.println(messageText);
                DingtalkUtils.pushInfoToDingding(messageText,
"30709c7b3f38101b853d1c50a211031d2ad358c1dca40e57d1c6f8435938f56c");
                break;
            // 服务未知异常
            case "UNKNOWN":
                log.info("发送 服务未知异常 的通知！");
                messageText = String.format(template, titleAlarm,
instance.getRegistration().getName(), event.getInstance(),
((InstanceStatusChangedEvent) event).getStatusInfo().getStatus(), "服务未
知异常通知",instance.getRegistration().getServiceUrl(), JSONObject.
toJSONString(instance.getStatusInfo().getDetails()));
                // 先输出信息在控制台
                DingtalkUtils.pushInfoToDingding(messageText,
"30709c7b3f38101b853d1c50a211031d2ad358c1dca40e57d1c6f8435938f56c");
```

```
                    break;
                default:
                    break;
            }
        } else {
            log.info("Instance {} ({}) {}", instance.
getRegistration().getName(), event.getInstance(), event.getType());
        }
    });
}
```

DingtalkNotifier 类继承了 AbstractStatusChangeNotifier，重写了 shouldNotify 和 doNotify 方法，当服务状态有变化时，会触发事件，通过 DingtalkUtils 将消息发送出去。

测试一下，在前面 3 个服务的基础上重启 passenger-api，钉钉通知如图 8-13 所示。

图 8-13　钉钉通知

8.6　小　　结

本章介绍了 Spring Boot Admin 组件的使用方法。比起通过直接看日志，通过 Spring Boot Admin 的界面观察程序的运行状态的方式更加方便和友好。本章详细介绍了监控界面的使用，以及在服务状态有变化时，如何及时通知开发人员，实现了邮件通知和钉钉通知两种方式，使得在服务发生故障时，能及时收到消息。

第 9 章 分布式锁解决方案

在实际开发的过程中,有的时候需要对并发进行控制。例如,当多个执行请求同时进入一个方法时,会出现数据不一致的问题。又如,电商业务中,在扣减库存后生成订单的逻辑中,需要先判断库存是否大于 0,如果大于 0,则再进行订单的创建。此时,如果多个请求(如 10 个)同时进来,假定此时库存是 2,那么每个请求对库存的判断都大于 0,这时候,10 个请求都会判断库存大于 0,就会创建 10 个订单,因此就会发生数据不一致的问题,但是前提库存只有 2 个,这就是常说的超卖问题。在这种场景下,就需要用锁来进行控制,防止多个请求同时进入方法中。提到锁的概念,当系统是单点时,通过 JVM 锁(如用 synchronized、lock)可以实现。在分布式项目中,一般情况下,项目为了避免单点故障,都会做集群。当系统做集群时,普通的 JVM 锁就会失效,因为集群中每个 JVM 进程拿到的是不同的锁。并且加锁会导致各种各样的问题,如死锁、容错、互斥、数据一致性等问题。本章就来学习如何制定一套完美的分布式锁解决方案。

9.1 业务场景

分布式锁的应用场景非常广,下面是几种特别常见的场景。
- ☑ 在外卖系统中,多个外卖员抢同一个用户的订单。
- ☑ 在电商系统中,多个用户购买同一件商品。
- ☑ 在网约车系统中,多个司机抢同一个乘客的订单。

这几种场景在工作中是经常会遇到的,如果能解决这几种场景的问题,当再遇到类似的场景问题时,解决方案基本都是通用的。

下面以秒杀抢购茅台酒为例(如 10 个人抢购 2 瓶茅台酒),来逐个递进学习分布式锁。抢购茅台酒的业务流程如下。

(1)查询茅台酒的库存。

（2）判断库存是否大于 0。
- ☑ 如果库存大于 0，执行购买操作，扣减库存，新增订单。
- ☑ 如果库存小于 0，提醒用户商品卖光了。

核心代码流程如图 9-1 所示。

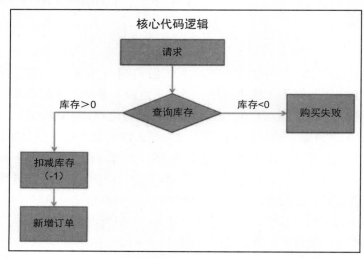

图 9-1　核心代码流程

核心业务 Java 代码如下所示。

```java
public boolean grab(int goodId, int userId) {
    TblInventory tblInventory = mapper.selectByPrimaryKey(goodId);
    try {
        Thread.sleep(2000);
    } catch (InterruptedException e) {
        e.printStackTrace();
    }
    // 获取库存值
    int num = tblInventory.getNum().intValue();
    System.out.println("num:"+num);
    if (num > 0) {
        // 扣减库存
        tblInventory.setNum(num-1);
        mapper.updateByPrimaryKeySelective(tblInventory);
```

```
    // 新增订单
    TblSeckillOrder order = new TblSeckillOrder();
    order.setOrderDescription("用户"+userId+"抢到了茅台");
    order.setOrderStatus(1);
    order.setUserId(userId);
    seckillOrderDao.insert(order);

    return true;
  }
  return false;
  }
}
```

接着准备数据库脚本。

创建库存表 tbl_inventory，代码如下。

```
CREATE TABLE `tbl_inventory` (
  `good_id` int(16) NOT NULL,
  `num` int(8) DEFAULT NULL,
  `update_time` timestamp NULL DEFAULT CURRENT_TIMESTAMP ON UPDATE CURRENT_TIMESTAMP,
  PRIMARY KEY (`good_id`)
) ENGINE=InnoDB DEFAULT CHARSET=utf8;
```

默认要抢购的商品 ID 是 1，提前在数据库预制一条数据。预制了 1 号商品，库存为 2。

```
INSERT INTO `tbl_inventory` VALUES ('1', '2', '2021-03-28 23:33:53');
```

创建订单表 tbl_seckill_order，代码如下。

```
CREATE TABLE `tbl_seckill_order` (
  `order_id` int(8) NOT NULL AUTO_INCREMENT,
  `order_status` int(8) DEFAULT NULL,
  `order_description` varchar(128) CHARACTER SET utf8 COLLATE utf8_general_ci DEFAULT NULL,
  `user_id` int(8) DEFAULT NULL,
  `update_time` timestamp NULL DEFAULT CURRENT_TIMESTAMP ON UPDATE CURRENT_TIMESTAMP,
  PRIMARY KEY (`order_id`)
) ENGINE=InnoDB AUTO_INCREMENT=434 DEFAULT CHARSET=utf8;
```

从数据库的角度来看，其业务逻辑是先扣减库存表 tbl_inventory 中的库存，然后在订单表 tbl_seckill_order 中新增一条订单数据。

9.2 单机 JVM 锁

9.2.1 系统架构与核心代码

接下来从单机版开始学习，系统的业务架构如图 9-2 所示。

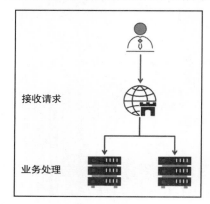

图 9-2 业务架构图

在业务架构图中将业务逻辑的处理分成以下两层。

（1）第一层，业务层。创建项目 order-api 用于接收用户的请求，通过调用其他服务的接口，来完成整个业务功能。

在这个服务中，可以通过 Ribbon 进行负载均衡，便于后面对集群的测试。这个服务主要是接收用户的请求，给用户暴露一个接口，代码如下。

```
@RestController
@RequestMapping("/grab")
public class GrabOrderController {

    @Autowired
    private RestTemplate restTemplate;
```

```
@GetMapping("/do/{orderId}")
public String grabMysql(@PathVariable("orderId") int orderId, int userId){

    // 电商秒杀
    String url = "http://service-order" + "/seckill/do/"+orderId+"?userId="+userId;

    restTemplate.getForEntity(url, String.class).getBody();
    return "成功";
    }
}
```

这个 Controller 提供的接口为 grab/do/1?userId=${userId}，这个是用户调用系统的入口。通过 RestTemplate 实现对服务层的调用（这个内容在服务调用环节学习过）。

（2）第二层，服务层。创建项目 service-order，这个服务提供核心的业务逻辑处理，例如 9.1 节中的核心代码，主要提供了如下接口。

```
@RestController
@RequestMapping("/seckill")
public class SeckillOrderController {

    @Autowired
    @Qualifier("seckillNoLockService")
    private SeckillGrabService grabService;

    @GetMapping("/do/{goodsId}")
    public String grabMysql(@PathVariable("goodsId") int goodsId, int userId){
        System.out.println("goodsId:"+goodsId+",userId:"+userId);
        grabService.grabOrder(goodsId,userId);
        return "";
    }
}
```

Controller 提供的接口为 seckill/do/{goodsId}?userId={userId}，负责接收业务层 order-api 的调用。

接口调用的服务代码如下。

```java
import com.online.taxi.order.service.GrabService;
import com.online.taxi.order.service.OrderService;
import com.online.taxi.order.service.SeckillGrabService;
import com.online.taxi.order.service.SeckillOrderService;
import org.springframework.beans.factory.annotation.Autowired;
import org.springframework.stereotype.Service;

@Service("seckillNoLockService")
public class SeckillNoLockServiceImpl implements SeckillGrabService {

    @Autowired
    SeckillOrderService seckillOrderService;

    @Override
    public String grabOrder(int goodsId, int userId) {
        try {
            System.out.println("用户:"+userId+" 执行秒杀");
            boolean b = seckillOrderService.grab(goodsId, userId);
            if(b) {
                System.out.println("用户:"+userId+" 抢单成功");
            }else {
                System.out.println("用户:"+userId+" 抢单失败");
            }

        } finally {
        }

        return "业务处理完成";
    }
}
```

在上面的 Service 层中调用核心的业务逻辑处理代码。

```java
import com.online.taxi.order.dao.TblInventoryDao;
import com.online.taxi.order.dao.TblOrderDao;
import com.online.taxi.order.dao.TblSeckillOrderDao;
import com.online.taxi.order.entity.TblInventory;
```

```java
import com.online.taxi.order.entity.TblOrder;
import com.online.taxi.order.entity.TblSeckillOrder;
import com.online.taxi.order.service.SeckillOrderService;
import org.springframework.beans.factory.annotation.Autowired;
import org.springframework.stereotype.Service;

@Service("seckillOrderService")
public class SeckillOrderServiceImpl implements SeckillOrderService {

    @Autowired
    private TblInventoryDao mapper;

    @Autowired
    private TblSeckillOrderDao seckillOrderDao;

    @Override
    public boolean grab(int goodId, int userId) {
        TblInventory tblInventory = mapper.selectByPrimaryKey(goodId);
        try {
            Thread.sleep(2000);
        } catch (InterruptedException e) {
            e.printStackTrace();
        }
        // 获取库存值
        int num = tblInventory.getNum().intValue();
        System.out.println("num:"+num);
        if (num > 0) {
            // 扣减库存
            tblInventory.setNum(num-1);
            mapper.updateByPrimaryKeySelective(tblInventory);

            // 新增订单
            TblSeckillOrder order = new TblSeckillOrder();
            order.setOrderDescription("用户"+userId+"抢到了茅台");
            order.setOrderStatus(1);
            order.setUserId(userId);
            seckillOrderDao.insert(order);
```

```
            return true;
        }
        return false;
    }
}
```

启动如下服务。

- ☑ eureka：注册中心，用于让 order-api 和 service-order 进行注册，可以让服务间通过服务名（虚拟主机名）进行调用。
- ☑ order-api：订单 API，这个是业务层，提供了 grab/do/1?userId=${userId}接口，用于接收用户的请求。
- ☑ service-api：订单服务，这个是服务层，提供了 seckill/do/{goodsId}?userId={userId} 接口，用于业务层 order-api 的调用。

9.2.2　JMeter 安装与配置

本节用 JMeter 工具来进行系统的压力测试。首先将介绍 JMeter 的安装与配置。

打开 JMeter 官网 https://jmeter.apache.org/，如图 9-3 所示。

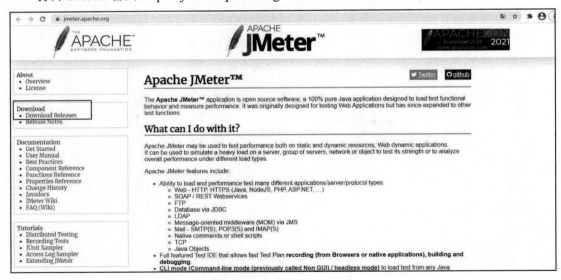

图 9-3　JMeter 官网

单击左侧的 Download Releases 链接，选择合适的版本进行下载，如图 9-4 所示。

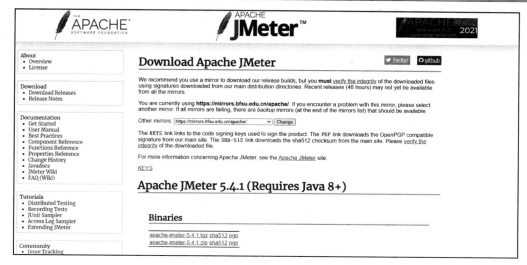

图 9-4 JMeter 下载

下载后得到一个 apache-jmeter-5.4.1.tgz 文件，解压即可。

文件解压后，双击 bin/jmeter.bat 运行，JMeter 的初始界面如图 9-5 所示。

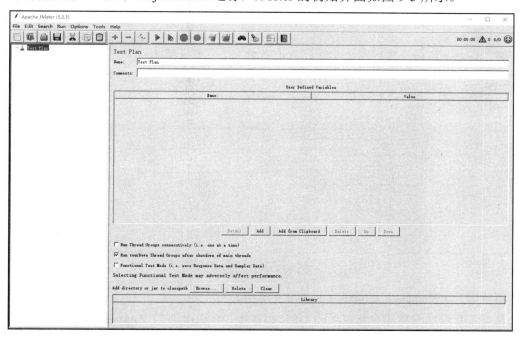

图 9-5 JMeter 初始界面

接下来创建线程组，如图 9-6 所示。

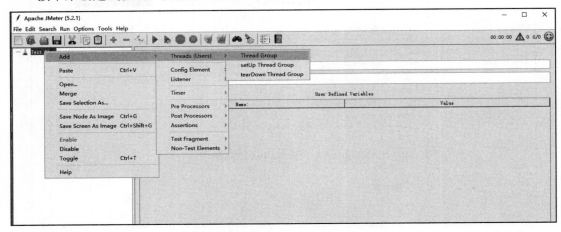

图 9-6　创建线程组

设置下列线程组参数，如图 9-7 所示。

- ☑　Number of Threads (users)：线程数量。设置为 10 个。
- ☑　Ramp-up period (seconds)：所有线程多次时间内启动。这里设置为 1，表示 1s 内同时启动。
- ☑　Loop Count：线程重复次数。这里设置为 1，表示循环一次。

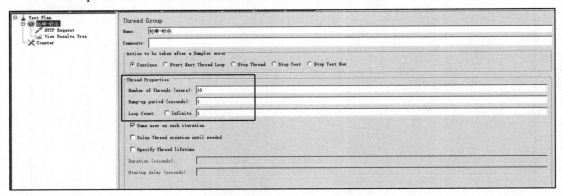

图 9-7　线程组参数

新增 HTTP Request，如图 9-8 和图 9-9 所示。

在图 9-9 所示的界面设置请求的 URL 和参数，注意此时的 Parameters。通过 userId 进行了一个参数变量的设置。userId 的值通过一个计数器来设置，如图 9-10 和图 9-11 所示。

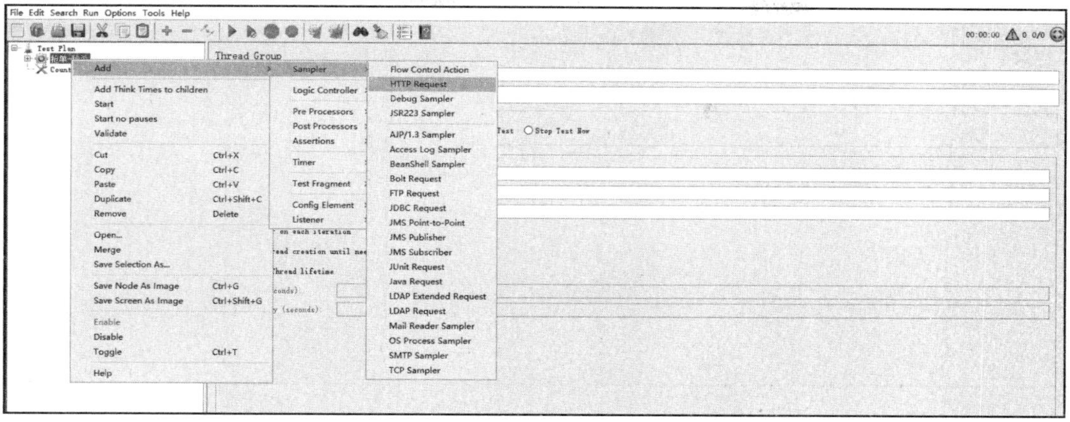

第9章 分布式锁解决方案

图 9-8 设置 HTTP Request

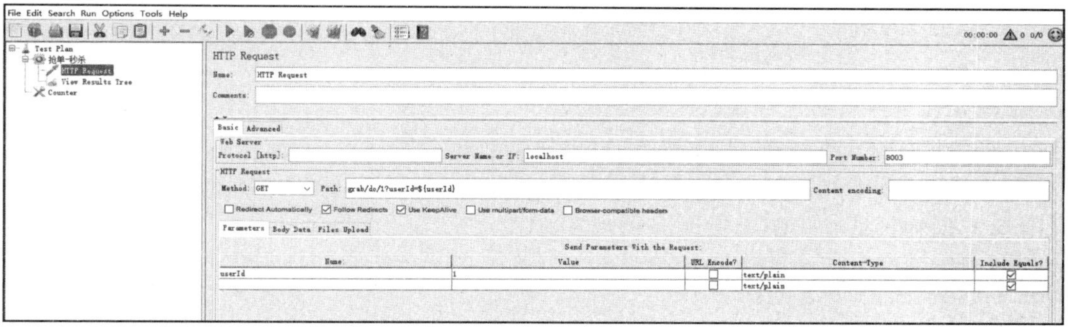

图 9-9 HTTP Request 详细设置

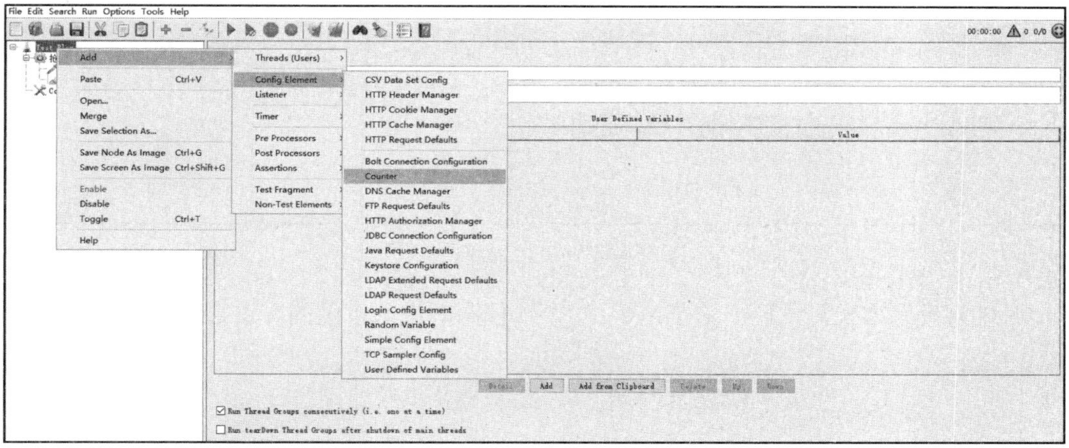

图 9-10 新建计数器

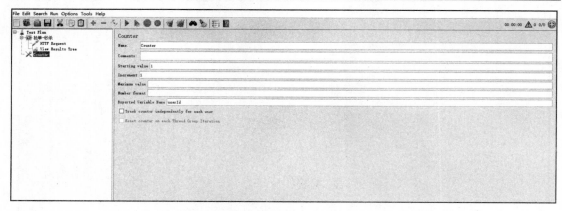

图 9-11　计数器详细设计

9.2.3　压力测试

在测试之前,我们先根据代码逻辑指定一个期望结果。现在模拟 10 个用户来抢购 2 个商品,期望的结果是库存从 2 变成 0,订单新增 2 条。

单击图 9-12 中的▶按钮,等待程序完成。运行结果是,库存从 2 变成了 1,订单新增了 10 条,这和预期结果不一致。

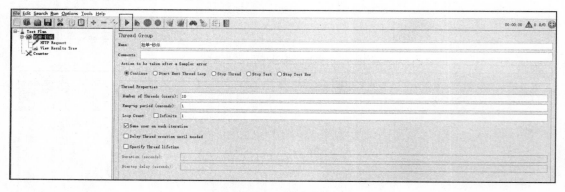

图 9-12　压测开始

通过分析原因我们发现,这是由于 10 个线程同时进入下面的方法导致的。

```
public boolean grab(int goodId, int userId) {
    TblInventory tblInventory = mapper.selectByPrimaryKey(goodId);
    try {
        Thread.sleep(2000);
```

```java
    } catch (InterruptedException e) {
        e.printStackTrace();
    }
    // 获取库存值
    int num = tblInventory.getNum().intValue();
    System.out.println("num:"+num);
    if (num > 0) {
        // 扣减库存
        tblInventory.setNum(num-1);
        mapper.updateByPrimaryKeySelective(tblInventory);

        // 新增订单
        TblSeckillOrder order = new TblSeckillOrder();
        order.setOrderDescription("用户"+userId+"抢到了茅台");
        order.setOrderStatus(1);
        order.setUserId(userId);
        seckillOrderDao.insert(order);

        return true;
    }
    return false;
}
```

10个线程都查询到库存为2，然后都进入了if(num>0)的循环，都执行了代码tblInventory.setNum(2-1)。所以库存结果为1。因为有10个线程进入了if(num>0)循环，10个线程执行了订单的insert操作，所以订单新增了10条，导致结果和预期不符。其实这就是在电商项目中经常说的**超卖问题**，即库存少了1个，结果却卖出去10个商品。

通过上面的分析找到了问题的原因，那解决办法也很简单，就是控制进入grab(int goodId, int userId)方法的线程。不能让10个线程同时进入，即只有当一个线程处理完时，另一个线程才能进入。这样就能避免上面所说的问题。解决方案是，在进入方法前进行同步的控制，代码如下。

```java
/**
 * 秒杀controller
 * @author 马士兵教育：晁鹏飞
 */
```

```java
@RestController
@RequestMapping("/seckill")
public class SeckillOrderController {

    // 无锁
    @Autowired
    // JVM 锁
    @Qualifier("seckillJvmLockService")
    private SeckillGrabService grabService;

    @GetMapping("/do/{goodsId}")
    public String grabMysql(@PathVariable("goodsId") int goodsId, int userId){
        System.out.println("goodsId:"+goodsId+",userId:"+userId);
        grabService.grabOrder(goodsId,userId);
        return "";
    }

}
```

其中，@Qualifier("seckillJvmLockService")对应的服务代码如下。

```java
import com.online.taxi.order.service.SeckillGrabService;
import com.online.taxi.order.service.SeckillOrderService;
import org.springframework.beans.factory.annotation.Autowired;
import org.springframework.stereotype.Service;

@Service("seckillJvmLockService")
public class SeckillJvmLockServiceImpl implements SeckillGrabService {

    @Autowired
    SeckillOrderService seckillOrderService;

    @Override
    public String grabOrder(int goodsId, int userId) {
        String lock = (goodsId+"");

        synchronized (lock.intern()) {
```

```
            try {
                System.out.println("用户:"+userId+" 执行秒杀");
                boolean b = seckillOrderService.grab(goodsId, userId);
                if(b) {
                    System.out.println("用户:"+userId+" 抢单成功");
                }else {
                    System.out.println("用户:"+userId+" 抢单失败");
                }
            } finally {
            }
        }
        return null;
    }
}
```

重启 service-order，将恢复库存为 2，清空订单表后再进行压力测试。发现结果和预期一致，即库存变成了 0，订单新增了 2 条。

9.2.4 单机 JVM 锁的问题

在实际生产中的服务都不是一台，因为从高可用的角度出发，要避免单点故障。一般都会给每个服务做集群，集群架构如图 9-13 所示。

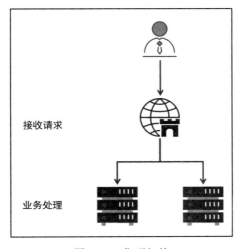

图 9-13 集群架构

为了启动两个程序，下面修改 application.yml 文件，设置 profiles 分别为 8004 和 8005。

```
spring:
  profiles: 8004
#服务端口
server:
  port: 8004
---
spring:
  profiles: 8005
#服务端口
server:
  port: 8005
```

在 IDEA 中，需要设置两个 Active profiles，分别如图 9-14 和图 9-15 所示。

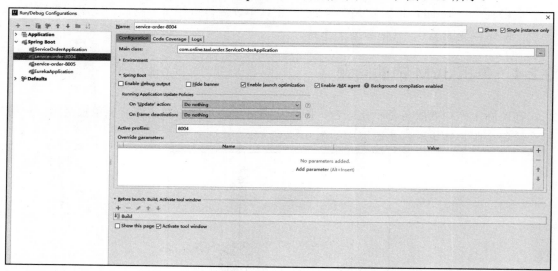

图 9-14　设置 Active profiles 为 8004

启动两个 service-order 服务，通过设置不同的 profiles（8004 和 8005）来完成。恢复一下数据库，将库存设置成 2，订单全部删除。预期结果如下。

☑　库存从 2 变成 0。
☑　订单新增 2 条。

第 9 章　分布式锁解决方案

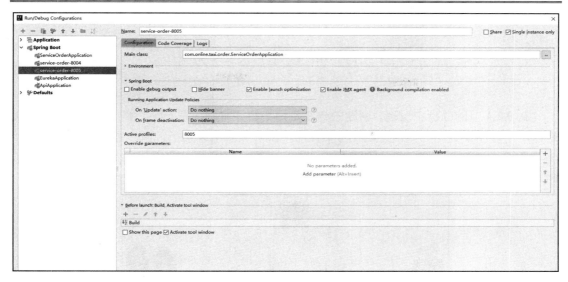

图 9-15　设置 Active profiles 为 8005

压力测试结束后，观察结果。实际结果如下。

☑　库存从 2 变成了 0。

☑　订单新增了 4 条。

但这与预期不符合。问题出在哪里呢？

分析一下代码，启动了两个 service-order（8004 和 8005）。通过 synchronized 加锁，程序加的锁分别是两个 JVM 进程中的锁，也就是它们（8004 和 8005）拿到的并不是同一把锁。因此导致两个程序都同时执行扣减库存和新增订单的操作，所以数据和预期不符，如图 9-16 所示。

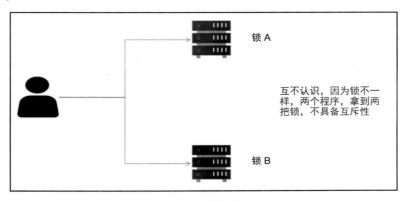

图 9-16　两个程序拿到两把锁

9.3 分布式锁思路分析

从 9.2.4 节的分析中发现,问题的核心是两个程序拿到了两把锁,解决办法也简单,就是让两个程序拿到同一把锁。但在每个程序中都加锁就不现实了,所以需要引入第三方。

从图 9-17 可以看出,引入第三方后,每个需要加锁的程序都去第三方拿锁,即当一个程序拿到锁后,另外一个程序再去拿锁时就拿不到了,因为第三方的锁只能被一个程序拿走。

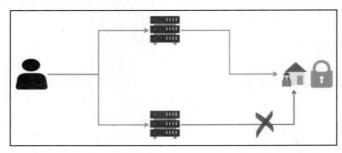

图 9-17 分布式锁思路

其实这就是分布式锁。

分布式锁是控制分布式系统或不同系统之间共同访问共享资源的一种锁实现,如果不同的系统或同一个系统的不同主机之间共享了某个资源时,往往需要互斥来防止彼此干扰来保证一致性。

分布式锁需要满足以下条件。

- ☑ 互斥性:在任意一个时刻,只有一个客户端持有锁。
- ☑ 无死锁:即便持有锁的客户端崩溃或者发生其他意外事件,锁仍然可以被其他服务获取。
- ☑ 容错:只要大部分第三方存储还活着,客户端就可以正常获取和释放锁。

以文件系统举例。例如,第三方是一块磁盘,有一个目录:"电商系统",在"电商系统"目录下,要锁定"商品 1"的 ID 时,就在"电商系统"的目录下,创建一个文件名为"商品 1"的文件,当前程序就拿到了锁。当其他业务程序也要锁定"电商系统"下的"商品 1"时,发现磁盘上已经有这个文件了,就不能创建"商品 1"这个文件了,也就拿不到锁。此时就解决了多个业务程序如何互斥加锁的问题。

当要释放锁时,删除自己创建的文件即可,也就是删除上面例子中的"商品 1"文件,其他程序又可以来这个目录下创建"商品 1"的文件了。文件目录分布式锁例子如图 9-18 所示。

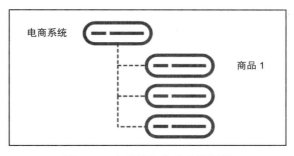

图 9-18　文件目录分布式锁例子

通过上面的分析可以得出一个结论，为了确保分布式锁的可用，需要满足以下 4 个条件。

- ☑ 互斥性：任意时刻只能有一个客户端获取锁，不能同时有两个客户端获取到锁。
- ☑ 安全性：锁只能被持有该锁的客户端删除，不能由其他客户端删除。
- ☑ 避免死锁：获取锁的客户端因为某些原因（如宕机等）而未能释放锁，其他客户端再也无法获取到该锁。
- ☑ 容错性：当部分节点（Redis 节点等）宕机时，客户端仍然能够获取锁和释放锁。

在生产中一般不用文件系统作为分布式锁的第三方组件，通常会采用 MySQL、Redis、Zookeeper 作为获取分布式的第三方。下面将分别讲解分步式锁的第三方组件。

9.4　MySQL 分布式锁

如果采用 MySQL 作为获取分布式锁的第三方组件，如何保证互斥性呢？类似于 9.3 节中提到的文件名不能重复的问题，我们很快想到主键不能重复，唯一索引不能重复。

那解决方案就很简单，创建一张表 tbl_distributed_lock。

```
CREATE TABLE `tbl_distributed_lock` (
 `goods_id` int(16) NOT NULL,
 `lock_start_time` datetime DEFAULT NULL ON UPDATE CURRENT_TIMESTAMP,
 `lock_end_time` datetime DEFAULT NULL ON UPDATE CURRENT_TIMESTAMP,
 PRIMARY KEY (`goods_id`)
) ENGINE=InnoDB DEFAULT CHARSET=utf8;
```

当需要抢购"商品 ID"为 1 的商品时，就向这张表中插入一条记录，其中 goods_id 是商品的 ID（1），当记录插入成功就拿到了锁。这时候当其他程序来抢这个商品时，由于主键不能重复，就不能再插入这条数据了，拿锁也就失败了。

释放锁也比较简单，就是删除这条记录。

如果在每次需要加锁的代码处都写这么一段加锁和释放锁的逻辑，是挺麻烦的。可以开发一个 MySQL 加锁和释放锁的工具类 MysqlLock。

```java
@Service
@Data
public class MysqlLock implements Lock {

    @Autowired
    private TblOrderLockDao mapper;
    private ThreadLocal<TblOrderLock> orderLockThreadLocal ;

    @Override
    public void lock() {
        // 1.尝试加锁
        if(tryLock()) {
            System.out.println("尝试加锁");
            return;
        }
        // 2.休眠
        try {
            Thread.sleep(10);
        } catch (InterruptedException e) {
            e.printStackTrace();
        }
        // 3.递归再次调用
        lock();
    }

    /**
     * 非阻塞式加锁，成功就成功，失败就失败，直接返回
     */
    @Override
    public boolean tryLock() {
        try {
```

```java
        TblOrderLock tblOrderLock = orderLockThreadLocal.get();
        mapper.insertSelective(tblOrderLock);
        System.out.println("加锁对象："+orderLockThreadLocal.get());
        return true;
    }catch (Exception e) {
        return false;
    }
}

@Override
public void unlock() {
    mapper.deleteByPrimaryKey(orderLockThreadLocal.get().getOrderId());
    System.out.println("解锁对象："+orderLockThreadLocal.get());
    orderLockThreadLocal.remove();
}

@Override
public void lockInterruptibly() throws InterruptedException {

}

@Override
public boolean tryLock(long time, TimeUnit unit) throws
InterruptedException {
    return false;
}

@Override
public Condition newCondition() {
    return null;
}

}
```

在需要使用加锁和释放锁的程序中引入它即可。

```java
@Autowired
private MysqlLock lock;
```

加锁的操作如下所示。

```
lock.lock();
```

释放锁的操作如下所示。

```
lock.unlock();
```

一般在一些小型系统中可以使用 MySQL 做分布式锁的第三方，虽然有能达到 QPS 为 10W 的 MySQL，但是需要投入的成本是巨大的。一般会从性能的角度分析，MySQL 是磁盘 IO 型数据库，如果为了提高性能，可以采用内存 IO 型的数据库，如 Redis。

9.5　Redis 分布式锁

当采用 Redis 做分布式锁时，如果要保证互斥性，需要使用下面的命令。

```
setnx key value
```

SETNX 命令的格式为 setnx key value。当且仅当 key 不存在时，将 key 的值设为 value。若给定的 key 已经存在，则 SETNX 不做任何动作。

SETNX 是 "SET if Not eXists"（如果不存在，则 SET）的简写。

当一个程序想要对"商品 ID"为 1 的商品进行加锁时，就会执行如下命令。

```
setnx 1 "值"
```

这样，当其他程序还想对"商品 ID"为 1 的商品加锁，通过 SETNX 命令就设置不成功了，因此也就拿不到锁了。上面是 Redis 加锁的核心命令，在整个 Redis 加锁过程中还有更多的注意事项。下面我们将一个一个来分析。

9.5.1　死锁问题

当在 Redis 中通过 SETNX 命令来加锁，当加锁后，锁永远存在于 Redis 中，这样其他程序来加锁时就永远拿不到锁了，就会造成死锁。所以要有一个释放锁的步骤来删除 key，即释放锁。

```
setnx key value
业务逻辑;
del key
```

来思考一个问题：当程序在执行业务逻辑时突然宕机了，还没有执行 del 操作，key 是会留在 Redis 中，还是会造成死锁？这个时候就需要给 key 设置过期时间了。可以使用 expire 命令来设置过期时间，命令如下。

```
setnx key value
expire key 10
业务逻辑；
del key
```

执行以上命令后，看似解决了 key 不过期的问题，但是仔细思考后再看，如果在执行完第一条命令 setnx key value 后，程序宕机了，并没有给 key 设置过期时间，还是会造成死锁。此时需要让 setnx 和设置过期时间是一个原子操作。Redis 支持以下命令。

```
set key value nx ex 10
```

关于上面的命令，有的读者喜欢用 lua 脚本来实现，这也是可以的。在 resource 下创建 luascript 目录，在里面编写 lua 脚本 lock-set.lua，代码如下。

```lua
--- 获取 key
local key = KEYS[1]
--- 获取 value
local val = KEYS[2]
--- 获取一个参数
local expire = ARGV[1]
--- 如果 redis 找不到这个 key 就去插入
if redis.call("get", key) == false then
    --- 如果插入成功，就去设置过期值
    if redis.call("set", key, val) then
        --- 由于 lua 脚本接收到参数都会转为 String，所以要转成数字类型才能比较
        if tonumber(expire) > 0 then
            --- 设置过期时间
            redis.call("expire", key, expire)
        end
        return true
    end
    return false
else
    return false
end
```

在程序中编写一个 Configuration。

```
@Configuration
public class LuaConfiguration {
    @Bean(name = "set")
    public DefaultRedisScript<Boolean> redisScript() {
        DefaultRedisScript<Boolean> redisScript = new DefaultRedisScript<>();
        redisScript.setScriptSource(new ResourceScriptSource(new ClassPathResource("luascript/lock-set.lua")));
        redisScript.setResultType(Boolean.class);
        return redisScript;
    }
}
```

在程序使用的时候直接注入即可。

```
@Resource(name = "set")
private DefaultRedisScript<Boolean> redisScript;

@Resource
private StringRedisTemplate stringRedisTemplate;
```

lua 的使用方式如下。

```
List<String> keys = Arrays.asList("testLua", "hello lua");
Boolean execute = stringRedisTemplate.execute(redisScript, keys, "100");
return null;
```

此时 List 中的"testLua"和"hello lua"分别对应于上面的 KEYS[1]和 KEYS[2]。
这样就彻底解决了在程序宕机时 key 不释放导致的死锁问题。

9.5.2 过期时间问题

当上面的 key 有了过期时间，接着思考：如果设置了 key 的有效期是 10s，结果程序执行了 15s，那么在最后 5s 时，key 已经过期了，如果有另外一个线程过来加锁，还是能加锁成功的，这样在最后 5s 就会有两个线程都拿到了锁，因此依然发生了前面的超卖问题。

还有一个能引起错乱的问题，当第二个线程正在执行业务时，第一个线程开始执行释放锁的代码，执行了 del key，把第二个线程的锁释放了。这样以此类推，线程二会释放线程三的锁，线程三会释放线程四的锁，因此会导致程序继续错乱。

上面这两个问题都是由于过期时间和程序的执行时间不匹配导致的，那如何让程序的执行时间和锁的过期时间相匹配呢？由于程序在运行时的精确时间无法预估，所以也无法精确地设置 key 的过期时间。此时引入一个概念看门狗（watch dog）。

假想这么一件事，小明去卫生间时，卫生间门上的锁是有有效期的，类似于 Redis 中 key 的过期时间，当小明开门要进入时，先给锁设置一个过期时间 1min，小明无法预估自己实际使用卫生间的时间，怕自己还没用完卫生间，锁过期后别人闯进来。这时聪明的小明牵来一条狗，告诉小狗，当主人在卫生间里到第 30s 时，你要主动将锁再延长 1min，如果主人一直不出来，小狗就每过 30s 延长一下锁的有效期到 1min。

这样就解决了程序没执行完但锁过期的问题。那在程序中的小狗会有个专业术语 watch dog，它其实就是和业务线程并行的一个后台线程。它的任务就是在程序执行期间，不停地延长锁的有效期，直到程序执行完成。

遗留问题：线程一删除线程二锁的问题，方法很简单，在加锁时给 key 设置一个独一无二的 value 值，这样在删除锁之前做一次判断。如果是自己加的锁，则释放，不是自己加的锁，则不释放。

删除锁也可以通过 lua 脚本实现。在 resource 下创建 luascript 目录，在里面编写一个 lua 脚本 lock-del.lua，代码如下。

```
if redis.call("get",KEYS[1])==ARGV[1] then
  return redis.call("del",KEYS[1])
else
  return 0
end
```

在程序中编写一个 Configuration。

```
@Configuration
public class LuaConfiguration {

    @Bean(name = "del")
    public DefaultRedisScript<Boolean> redisScriptDel() {
        DefaultRedisScript<Boolean> redisScript = new DefaultRedisScript<>();
        redisScript.setScriptSource(new ResourceScriptSource(new ClassPathResource("luascript/lock-del.lua")));
        redisScript.setResultType(Boolean.class);
```

```
        return redisScript;
    }
}
```

在程序使用的时候直接注入即可。

```
List<String> keys = Arrays.asList("testLua");
Boolean execute = stringRedisTemplate.execute(redisScriptDel, keys,
"hello lua");
return null;
```

list 中的"testLua"对应于 lua 中的 KEYS[1]。

9.5.3 Redisson 框架使用

通过上面的分析发现，需要考虑的问题还挺多，如果要实现起来会很复杂，不过别担心，已经有人封装好了一个框架 Redisson，用它做分布式锁会方便很多。

首先，在 pom 文件中添加依赖。

```
<dependency>
    <groupId>org.redisson</groupId>
    <artifactId>redisson</artifactId>
    <version>3.3.2</version>
</dependency>
```

然后，配置 Redis，代码如下。

```
import org.redisson.Redisson;
import org.redisson.api.RedissonClient;
import org.redisson.config.Config;
import org.springframework.beans.factory.annotation.Autowired;
import org.springframework.boot.autoconfigure.condition.
ConditionalOnMissingBean;
import org.springframework.context.annotation.Bean;
import org.springframework.context.annotation.Primary;
import org.springframework.data.redis.connection.RedisConnectionFactory;
import org.springframework.data.redis.core.StringRedisTemplate;
import org.springframework.stereotype.Component;

/**
```

```
 * Redisson 配置类
 * @author 晁鹏飞
 */
@Component
public class RedisConfig {

    @Autowired
    RedisSentinelProperties properties;

    /**
     * 单个 redisson
     * @return
     */
    @Bean
    public RedissonClient redissonClient() {
        Config config = new Config();
        config.useSingleServer().setAddress("127.0.0.1:6379").setDatabase(0);

        return Redisson.create(config);
    }

}
```

接着，在代码中注入 RedissonClient。

```
@Autowired
RedissonClient redissonClient;
```

业务代码中的用法如下。

```
@Override
public String grabOrder(int goodsId, int userId) {
    // 生成 key
    String lock = "goodsId_"+(goodsId+"");

    RLock rlock = redissonClient.getLock(lock.intern());

    try {
```

```
        // 此代码默认设置key超时时间30s，过10s，再延时
        rlock.lock();

        boolean b = seckillOrderService.grab(goodsId, userId);
        if(b) {
            System.out.println("用户:"+userId+" 抢单成功");
        }else {
            System.out.println("用户:"+userId+" 抢单失败");
        }

    } finally {
        rlock.unlock();
    }
    return null;
}
```

9.5.4 Redis 单节点问题

通过上面的学习，基本已经把 Redis 做分布式锁的常用的注意事项阐述清楚了。但是如果只用一台 Redis 的话，还会有 Redis 的单点故障。当 Redis 发生故障时，所有程序都不能加锁，此时对业务的影响将是灾难性的。

为了避免单点故障，可以做 Redis 集群，常用的是一主二从三哨兵。但是有这样一种场景：线程一去 master 加锁时，setnx 执行成功，线程一拿到了锁。由于 master 和 slave 之间的数据同步不是实时的，中间有个时间差，而此时 master 发生故障，那么 Redis 集群会将 slave 选举为 master，不巧的是，就在此时，线程二也来拿锁，由于最开始的 master 没有将数据同步过来，所以新的 master（原来的 slave）中是没有原来加锁的 key 的，这样一来，线程二也能拿到锁。此时，又发生了一件事，线程一和线程二同时拿到了同一把锁。

所以为了避免单点故障，需要从以下两个角度出发。

- ☑ Redis 要用多台。
- ☑ 多台 Redis 之间不能有数据同步。

此时诞生了一种方案——红锁。

9.5.5 红锁

红锁，其实就是多台独立的 Redis 来做获取锁的第三方。它本来是 Redis Lock，是 Redis

的作者提出的一种分布式锁的解决方案，简写为 Red Lock，所以叫红锁。红锁一般用 N（奇数）台 Redis 做加锁的 Redis。

红锁的加锁步骤如下。

（1）获取当前 Unix 时间，以 ms 为单位。

（2）轮流用相同的 key 和随机值在 N 个节点上请求锁，在这一步里，客户端在每个 master 上请求锁时，会有一个和总的锁释放时间相比小得多的超时时间。例如，如果锁的自动释放时间是 10s，那每个节点锁请求的超时时间可能是 5~50ms，这可以防止一个客户端在某个宕掉的 master 节点上阻塞过长时间，如果一个 master 节点不可用了，应该尽快尝试下一个 master 节点。

（3）客户端计算第二步中获取锁所花的时间，只有当客户端在大多数 master 节点上成功获取了锁（如果有 5 个节点，超过 3 台即可），而且总共消耗的时间不超过锁的释放时间，这个锁就认为是获取成功了。

（4）如果锁获取成功了，那现在锁自动释放时间就是最初的锁释放时间减去之前获取锁所消耗的时间。

（5）如果锁获取失败了，不管是因为获取成功的锁不超过一半（N/2+1）还是因为总消耗时间超过了锁释放时间，客户端都会到每个 master 节点上释放锁，即便是那些它认为没有获取成功的锁。

红锁释放锁的步骤就简单多了，只需要在所有节点都释放锁就行，不管之前有没有在该节点获取过锁。

红锁看似完美，既避免了 Redis 的单点故障，也避免了 Redis 集群中 master 和 slave 同步的问题。但是它还有一个问题。大家设想这样一个场景：有 5 台 Redis（分别为 1 号、2 号、3 号、4 号、5 号）作为分布式锁加锁的中间件。有一个线程 A，向 1 号、2 号、3 号 Redis 设置 key 成功，那么此时，线程 A 加锁成功，3 号 Redis 突然由于机器故障不能运行了，并且数据丢失了，运维同事迅速部署了一台 3 号 Redis（里面没有数据）启动，此时，线程 B 正好来加锁，那么它可以在 3 号、4 号、5 号这 3 台 Redis 上加锁成功。此时线程 A 和线程 B 在同一个业务执行期间，将获取到同一把锁，这将又产生了前面多个程序同时执行一个方法的问题。其实这个问题的解决方法也很简单，3 号 Redis 有故障，延迟启动即可，要么等线程 A 的业务执行完再启动，要么等 1 号和 2 号 Redis 中的 key 过期才启动。这样就不会有两个线程同时拿到锁来执行业务的情况了。

红锁的加锁步骤比较麻烦，但是有一个框架实现了它，因此我们直接使用该框架即可。

首先，在 pom 中添加依赖。

```
<dependency>
    <groupId>org.redisson</groupId>
```

```
    <artifactId>redisson</artifactId>
    <version>3.3.2</version>
</dependency>
```

然后，配置 Redis，代码如下所示。

```java
import org.redisson.Redisson;
import org.redisson.api.RedissonClient;
import org.redisson.config.Config;
import org.springframework.beans.factory.annotation.Autowired;
import org.springframework.boot.autoconfigure.condition.ConditionalOnMissingBean;
import org.springframework.context.annotation.Bean;
import org.springframework.context.annotation.Primary;
import org.springframework.data.redis.connection.RedisConnectionFactory;
import org.springframework.data.redis.core.StringRedisTemplate;
import org.springframework.stereotype.Component;

/**
 * 红锁配置类
 * @author 晁鹏飞
 */
@Component
public class RedisConfig {

    @Autowired
    RedisSentinelProperties properties;

    @Bean(name = "redissonRed1")
    @Primary
    public RedissonClient redissonRed1(){
        Config config = new Config();
        config.useSingleServer().setAddress("127.0.0.1:6379").setDatabase(0);
        return Redisson.create(config);
    }
    @Bean(name = "redissonRed2")
    public RedissonClient redissonRed2(){
```

```java
        Config config = new Config();
        config.useSingleServer().setAddress("127.0.0.1:6380").setDatabase(0);
        return Redisson.create(config);
    }
    @Bean(name = "redissonRed3")
    public RedissonClient redissonRed3(){
        Config config = new Config();
        config.useSingleServer().setAddress("127.0.0.1:6381").setDatabase(0);
        return Redisson.create(config);
    }
    @Bean(name = "redissonRed4")
    public RedissonClient redissonRed4(){
        Config config = new Config();
        config.useSingleServer().setAddress("127.0.0.1:6382").setDatabase(0);
        return Redisson.create(config);
    }
    @Bean(name = "redissonRed5")
    public RedissonClient redissonRed5(){
        Config config = new Config();
        config.useSingleServer().setAddress("127.0.0.1:6383").setDatabase(0);
        return Redisson.create(config);
    }
    // 以上为红锁

}
```

上面代码中的 5 个 RedissonClient，共配置了 5 台 Redis 信息。

在业务代码中的使用方法如下。

```java
import com.online.taxi.order.service.SeckillGrabService;
import com.online.taxi.order.service.SeckillOrderService;
import org.redisson.RedissonRedLock;
import org.redisson.api.RLock;
import org.redisson.api.RedissonClient;
import org.springframework.beans.factory.annotation.Autowired;
```

```java
import org.springframework.beans.factory.annotation.Qualifier;
import org.springframework.stereotype.Service;

import java.util.concurrent.TimeUnit;

/**
 * @author 马士兵教育：晁鹏飞
 */
@Service("seckillRedisRedissonRedLockLockService")
public class SeckillRedisRedissonRedLockLockServiceImpl implements SeckillGrabService {

    // 红锁
    @Autowired
    @Qualifier("redissonRed1")
    private RedissonClient redissonRed1;
    @Autowired
    @Qualifier("redissonRed2")
    private RedissonClient redissonRed2;
    @Autowired
    @Qualifier("redissonRed3")
    private RedissonClient redissonRed3;
    @Autowired
    @Qualifier("redissonRed4")
    private RedissonClient redissonRed4;
    @Autowired
    @Qualifier("redissonRed5")
    private RedissonClient redissonRed5;

    @Autowired
    SeckillOrderService seckillOrderService;

    @Override
    public String grabOrder(int goodsId, int userId) {
        System.out.println("红锁实现类");
        // 生成 key
        String lockKey = ("goodsId_" + goodsId).intern();
```

```
        // 红锁 redisson
        RLock rLock1 = redissonRed1.getLock(lockKey);
        RLock rLock2 = redissonRed2.getLock(lockKey);
        RLock rLock3 = redissonRed3.getLock(lockKey);
        RLock rLock4 = redissonRed4.getLock(lockKey);
        RLock rLock5 = redissonRed5.getLock(lockKey);
        RedissonRedLock rLock = new RedissonRedLock(rLock1,rLock2,rLock3,
rLock4,rLock5);

        try {
            // 红锁
            rLock.lock();
            boolean b = seckillOrderService.grab(goodsId, userId);
            if(b) {
                System.out.println("用户:"+userId+" 抢单成功");
            }else {
                System.out.println("用户:"+userId+" 抢单失败");
            }
        } finally {
            rLock.unlock();
        }
        return null;
    }
}
```

分析以上代码可以发现，首先注入了上一步骤中的 5 个 RedissonClient，5 个 RedissonClient 调用 getLock 方法得到 5 个 RLock，再利用 5 个 RLock 通过 new RedissonRedLock（5 个 RLock）方法得到一个 RedissonRedLock。利用 RedissonRedLock 就可以进行加锁（rLock.lock()）、释放锁（rLock.unlock()）的操作。

红锁也默认继承了 watch dog 的功能。可以通过在程序中设置睡眠时间来检测。例如，在红锁加锁后让程序睡眠 5min。

```
try {
  TimeUnit.MINUTES.sleep(5);
  } catch (InterruptedException e) {
  e.printStackTrace();
}
```

去 Redis 查看 5 个 key 的情况发现，key 的有效期是 30s，当 TTL 降低到 20s 时，会自动将 key 的有效期变成 30s。其实现的代码在 org.redisson 包下类 RedissonLock 中。

```java
private void scheduleExpirationRenewal(final long threadId) {
    if (expirationRenewalMap.containsKey(getEntryName())) {
        return;
    }

    Timeout task = commandExecutor.getConnectionManager().newTimeout(new TimerTask() {
        @Override
        public void run(Timeout timeout) throws Exception {

            RFuture<Boolean> future = commandExecutor.evalWriteAsync(getName(), LongCodec.INSTANCE, RedisCommands.EVAL_BOOLEAN,
                    "if (redis.call('hexists', KEYS[1], ARGV[2]) == 1) then " +
                        "redis.call('pexpire', KEYS[1], ARGV[1]); " +
                        "return 1; " +
                    "end; " +
                    "return 0;",
                      Collections.<Object>singletonList(getName()), internalLockLeaseTime, getLockName(threadId));

            future.addListener(new FutureListener<Boolean>() {
                @Override
                public void operationComplete(Future<Boolean> future) throws Exception {
                    expirationRenewalMap.remove(getEntryName());
                    if (!future.isSuccess()) {
                        log.error("Can't update lock " + getName() + " expiration", future.cause());
                        return;
                    }

                    if (future.getNow()) {
                        // 重新调度
                        scheduleExpirationRenewal(threadId);
                    }
```

```
            }
        });
    }
}, internalLockLeaseTime / 3, TimeUnit.MILLISECONDS);

if (expirationRenewalMap.putIfAbsent(getEntryName(), task) != null) {
    task.cancel();
}
```

看下面这一行代码。

```
Timeout task = commandExecutor.getConnectionManager().newTimeout(new
TimerTask(),internalLockLeaseTime / 3,TimeUnit.MILLISECONDS)
```

即程序的延迟执行时间是 30/3s（也就是 10s），当 key 的有效期过去 10s 时，程序将 key 进行延期。

9.5.6 Redis 做分布式锁的终极问题

其实用 Redis 并不能完美地解决 Java 程序加分布式锁的需求。大家假设这种场景，当进程 A 从 Redis 中刚刚拿到了锁，此时系统进行完整垃圾回收（full gc），进程全部停顿（STW，stop the world），如果持续的时间比较久，导致 watch dog 没有去续期，结果 Redis 中的 key 过期了。这时候，另外一个进程 B 正好有加锁需求，它完全可以在 Redis 中拿到锁（因为进程 A 加锁设置的 key 过期了）。那么此时，进程 A 刚好从 STW 中恢复过来，即又出现了两个程序都拿到了锁，都可以执行业务逻辑的问题。

这个问题用 JVM 和 Redis 无法解决。除非更改 JVM，让它不进行完整垃圾回收，或不用 Redis，此时 Redis 的 key 在没有续期的情况下不会过期。这种情况可以通过 Zookeeper 结合 MySQL 乐观锁的方式解决。

9.6 Zookeeper 分布式锁

若用 Zookeeper 做分布式锁的第三方组件，我们得先了解 Zookeeper 是什么。Zookeeper，顾名思义就是动物园管理员，用来管理大象（hadoop）、蜜蜂（hive）、小猪（pig）。它提供了配置管理、名字服务、分布式锁和集群管理。

9.6.1 Zookeeper 节点类型

先来查看 Zookeeper 的结点概念，Zookeeper 的数据存储结构就像一棵树，这棵树由结点组成，这种结点叫作 Znode。

Znode 分为以下 4 种类型。

- ☑ 持久结点（PERSISTENT）：默认的结点类型。当创建结点的客户端与 Zookeeper 断开连接后，该结点依旧存在。
- ☑ 持久结点顺序结点（PERSISTENT_SEQUENTIAL）：所谓顺序结点，就是在创建结点时，Zookeeper 根据创建的时间顺序给该结点名称进行编号。例如，在商品 1 结点下会有结点"商品 id-100001""商品 id-100002""商品 id-100003"等结点，如图 9-19 所示。

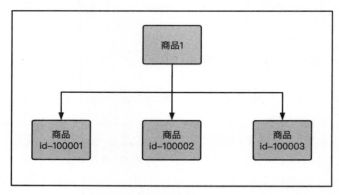

图 9-19 持久顺序节点

- ☑ 临时结点（EPHEMERAL）：和持久结点相反，当创建结点的客户端与 Zookeeper 断开连接后，临时结点会被删除。
- ☑ 临时顺序结点（EPHEMERAL_SEQUENTIAL）：临时顺序结点结合了临时结点和顺序结点的特点，在创建结点时，Zookeeper 根据创建的时间顺序给该结点名称进行编号。当创建结点的客户端与 Zookeeper 断开连接后，临时结点会被删除。

9.6.2 Zookeeper 分布式锁原理

分布式锁是控制分布式系统之间同步访问共享资源的一种方式。下面分别从获取锁和释放锁两个方面介绍 Zookeeper 的分布式锁原理。

1. 获取锁

在 Zookeeper 中创建一个持久结点"商品 1"。当第一个客户端 1 想要获得锁时,需要在"商品 1"这个结点下面创建一个临时顺序节点"商品 id-100001"。然后客户端 1 在"商品 1"下面查询所有的临时顺序结点并排序,判断自己所创建的结点是不是序号最小的一个,如果是,则成功获取锁。

此时,如果有其他客户端 2 再来获取锁,则在结点"商品 1"下面创建一个临时顺序结点"商品 id-100002",客户端 2 查找结点"商品 1"下面所有的临时顺序结点并排序,判断自己的结点"商品 id-100002"是不是最小的那一个,如果不是,客户端 2 向比它靠前的结点"商品 id-100001"注册 Watcher,用于监听"商品 id-100001"的状态,这表示客户端 2 抢锁失败,进入了等待状态。

这时,如果有其他客户端 3 再来获取锁,则在结点"商品 1"下面创建一个临时顺序结点"商品 id-100003",客户端 3 查找结点"商品 1"下面所有的临时顺序结点并排序,判断自己的结点"商品 id-100003"是不是最小的那一个,如果不是,客户端 3 向比它靠前的结点"商品 id-100002"注册 Watcher,用于监听"商品 id-100002"的状态,这表示客户端 3 抢锁失败,进入了等待状态,以此类推。

这样一来,客户端 1 拿到了锁,客户端 2 监听客户端 1,客户端 3 监听客户端 2,这就形成了一个等待队列。

2. 释放锁

释放锁有以下两种情况。
- ☑ 当任务完成时,客户端 1 会显示调用删除结点"商品 id-100001"的命令。
- ☑ 在执行任务过程中,客户端崩溃。根据临时结点的特性,Zookeeper 与客户端相关联的结点也会自动随之删除。利用这个特性就避免了死锁。

由于客户端 2 一直监听着客户端 1 的存在状态,当结点"商品 id-100001"被删除时,客户端 2 会立刻收到通知,这时客户端 2 会查询它对应的结点是不是结点"商品 1"下面顺序最小的结点,如果是,客户端 2 顺利拿到了锁。同理,当客户端 2 也会因为任务完成或者结点崩溃而删除了结点"商品 id-100002",则客户端 3 就会收到通知,进行结点大小的比较,如果结点最小,进而拿到锁。

9.6.3 Zookeeper 结合 MySQL 乐观锁

在 Redis 分布式锁章节中知道由于有 JVM 的 full gc 引起 STW,导致 watch dog 没有去

Redis 续期，而引发的锁失效的问题。可以通过 Zookeeper 结合 MySQL 乐观锁来解决。

可以利用 Zookeeper 结点顺序的特性，当进程 1 拿到锁时，将 Zookeeper 结点的序号 0 保存在程序中，并且在 MySQL 中存储下来，当最终操作业务的时候，通过 where 条件判断 MySQL 中的字段序号是否是 0，如果是，则更新，不是就回滚。

这么一来，即使进程 1 发生 STW，链接断开导致结点删除，此时进程 2 创建了结点，拿到了序号 1，并且在程序中保存了序号 1，重要的一点，更新了 MySQL 中的序号。如果进程 1 从 stop the world 中恢复过来去执行业务的时候，发现 MySQL 中的序号已经从序号 0 变成了序号 1，则进程 1 回滚，不影响数据的一致性。

9.6.4　Zookeeper 分布式锁代码实现

下面使用 curator 框架来实现，步骤如下。

（1）首先，修改 pom 文件，引入依赖。

```
<!-- Zookeeper 客户端 -->
<dependency>
    <groupId>org.apache.curator</groupId>
    <artifactId>curator-recipes</artifactId>
    <version>4.0.0</version>
    <exclusions>
        <exclusion>
            <groupId>org.apache.zookeeper</groupId>
            <artifactId>zookeeper</artifactId>
        </exclusion>
    </exclusions>
</dependency>

<dependency>
    <groupId>org.apache.zookeeper</groupId>
    <artifactId>zookeeper</artifactId>
    <version>3.4.14</version>
</dependency>
```

（2）创建 CuratorFramework。

```
/**
 * Zookeeper 客户端对象
```

```java
 * Curator Apache 是访问 Zookeeper 的工具包,既封装了这些低级别操作,也提供一些高级
服务,如分布式锁、领导选取
 * @return
 */
@Bean
public CuratorFramework curatorFramework(){
    // ExponentialBackoffRetry 是种重连策略,每次重连的间隔会越来越长,1000ms 是初
始化的间隔时间,3 代表尝试重连的次数
    ExponentialBackoffRetry retry = new ExponentialBackoffRetry(1000, 3);
    // 创建 client
    CuratorFramework curatorFramework = CuratorFrameworkFactory.newClient
("localhost:2181", retry);
    // 添加 watched 监听器
    curatorFramework.getCuratorListenable().addListener(new 
MyCuratorListener());
    curatorFramework.start();
    return curatorFramework;
}
```

(3)在业务代码中引入 CuratorFramework。

```java
/**
 * curator 中 zk 客户端对象
 */
@Autowired
private CuratorFramework client;
// 加锁业务使用
@Override
public String grabOrder(int goodsId, int userId) {
    // 抢锁路径,同一个锁路径需一致
    String lockPath = "/order/"+goodsId;

    // 创建分布式锁
    InterProcessMutex lock = new InterProcessMutex(client, lockPath);
    try {
        // 获取锁资源
        if (lock.acquire(10, TimeUnit.HOURS)) {
```

```
            System.out.println("用户:"+userId+" 执行秒杀");

            boolean b = seckillOrderService.grab(goodsId, userId);
            if(b) {
                System.out.println("用户:"+userId+" 抢单成功");
            }else {
                System.out.println("用户:"+userId+" 抢单失败");
            }
        }
    } catch (Exception e) {
        e.printStackTrace();
    } finally {
        try {
            lock.release();
            System.out.println("释放资源");
        } catch (Exception e) {
            e.printStackTrace();
        }
    }

    return null;
}
```

9.7 小　　结

  本节从一个业务场景出发，通过一个秒杀茅台酒的例子，介绍了从单机锁到分布式锁的解决方案，并讲解了它们的原理和实现。同时还通过介绍几个常用的中间组件MySQL、Redis和Zookeeper作为第三方，实现了分布式锁，并有实际的代码示例。详细地剖析了分布式中常遇到的死锁问题、单点问题。通过本节的学习，读者可以达到在生产环境中使用分布式锁的水平。

第 10 章 分布式事务解决方案

在实际工作中是需要对数据的一致性进行保证的，有一些数据持久化的操作必须是原子性。例如，在银行转账业务中，张三向李四转账 100 元，要保证张三账户中减少 100 元和李四账户中增加 100 元是一个原子操作，这就是所说的事务。提到事务，首先想到的是数据库事务。事务就是用户定义的一系列数据库操作，这些操作可以视为一个完整的逻辑处理工作单元，要么全部执行，要么全部不执行，它们是不可分割的工作单元。那么在微服务系统中，完成一个业务逻辑有时需要调用不同的服务来完成，而不同服务分布在不同机器上，甚至每个服务都有各自独立的数据库，这样就不能用传统的数据库事务来进行保证了，因为都不在一个数据库中了。那么在不同服务，甚至不同数据库中，如何保证它们之间的事务呢？这时就需要用到分布式事务的解决方案。

10.1 分布式事务业务场景

事务需要满足如下（ACID）特性。
- ☑ 原子性（atomicity），可以理解为一个事务内的所有操作要么都执行，要么都不执行。
- ☑ 一致性（consistency），可以理解为数据是满足完整性约束的，也就是不会存在中间状态的数据，例如，你的账户上有 400 元，我的账户上有 100 元，你给我转 200 元，此时你的账户上的钱应该是 200 元，我的账户上的钱应该是 300 元，不会存在我的账户上钱增加了，而你的账户上钱没扣减的中间状态。
- ☑ 隔离性（isolation），指的是多个事务并发执行时不会互相干扰，即一个事务内部的数据对于其他事务来说是隔离的。
- ☑ 持久性（durability），指的是一个事务完成后数据就被永远保存了下来，之后的其他操作或故障都不会对事务的结果产生影响。

其实事务的使用很简单，只需执行以下命令。

```
begin transaction;
// 执行业务;
commit/rollback;
```

commit 表示事务的提交操作,表示该事务的结束,此时将事务中处理的数据刷到磁盘中物理数据库磁盘中去。rollback 表示事务的回滚操作,若事务异常结束,此时将事务中已经执行的操作退回到原来的状态。

上面描述的是在单库的情况,事务要满足的条件以及操作方法。在微服务架构下,随着服务和数据库的拆分,就引入了分布式事务场景,如图 10-1 所示。

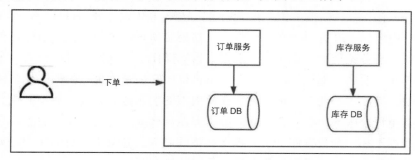

图 10-1 分布式事务场景

用户在电商业务中下单,首先调用库存服务扣减库存,然后调用订单服务新增订单,在单机事务的解决方案中,库存服务和订单服务是两个服务,它们也都有两个数据库,无法通过单机事务来保证,例如,库存服务不知道订单服务的执行结果,有可能会遇到库存扣减了,结果订单却没有新增的情况,这样就造成了数据的不一致。这就是分布式事务的业务场景。

下面将通过一段实际的业务代码来学习分布式事务。用户下单需要扣减库存,然后新增订单。这时的服务调用的链路是这样的:订单服务接收用户的请求,订单服务调用库存服务来完成库存的扣减,扣减库存成功后订单服务新增订单,分布式事务业务流程如图 10-2 所示。

分布式事务的操作步骤如下。

(1)创建表结构。

对于订单库,订单表的表结构如下。

```
CREATE TABLE `tbl_order` (
  `order_id` int(16) NOT NULL AUTO_INCREMENT COMMENT '订单Id',
  `goods_id` int(16) DEFAULT NULL COMMENT '商品ID',
  `buyer` varchar(32) CHARACTER SET utf8 COLLATE utf8_general_ci DEFAULT NULL COMMENT '买家',
  `update_time` timestamp NULL DEFAULT CURRENT_TIMESTAMP ON UPDATE
```

```
CURRENT_TIMESTAMP COMMENT '更新时间',
  PRIMARY KEY (`order_id`)
) ENGINE=InnoDB AUTO_INCREMENT=50 DEFAULT CHARSET=utf8;
```

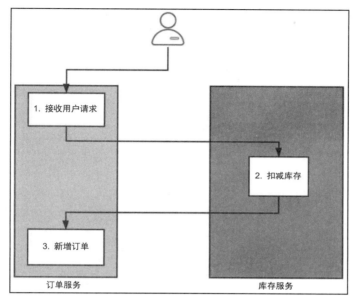

图 10-2　分布式事务业务流程

对于库存库，库存表的表结构如下。

```
CREATE TABLE `tbl_inventory` (
  `good_id` int(16) NOT NULL COMMENT '商品ID',
  `num` int(8) DEFAULT NULL COMMENT '库存数量',
  `update_time` timestamp NULL DEFAULT CURRENT_TIMESTAMP ON UPDATE
CURRENT_TIMESTAMP COMMENT '更新时间',
  PRIMARY KEY (`good_id`)
) ENGINE=InnoDB DEFAULT CHARSET=utf8;
```

（2）准备注册中心。

启动项目 Eureka 作为微服务注册中心（在 2.3 节中已经介绍过，在此不再赘述）。

（3）编写订单服务核心逻辑，代码如下。

```
@Transactional(rollbackFor = Exception.class)
public String addOrder(Integer goodsId) {
    restTemplate.getForEntity("http://inventory/reduce?goodsId="+
```

```
goodsId, null);

    TblOrder tblOrder = new TblOrder();
    tblOrder.setOrderId(1);
        tblOrder.setGoodsId(goodsId);
    tblOrder.setBuyer("买家-晁鹏飞");

    tblOrderDao.insert(tblOrder);

    return "";
}
```

上述代码中 restTemplate.getForEntity("http://inventory/reduce?goodsId="+goodsId, null); 实现了调用库存的服务。

（4）编写库存服务核心逻辑，代码如下。

```
@Transactional
public String reduce(int goodId) {
    TblInventory tblInventory = tblInventoryDao.selectByPrimaryKey(goodId);
    tblInventory.setNum(tblInventory.getNum()-1);

    tblInventoryDao.updateByPrimaryKey(tblInventory);

    return "";
}
```

上述代码进行了库存减 1 的操作。

（5）测试分布式事务。

通过 Postman 请求地址 localhost:2001/order-add?goodsId=1。

- ☑ 正常情况：库存表数量减 1，订单表数量加 1。
- ☑ 异常情况：在订单服务中手写一个异常 System.out.println(1/0);，结果库存表减了 1，而订单表并没有新增。

10.2 分布式事务思路分析

通过上面的现象得知，在一个请求的调用链路中，如果跨多个服务或者多个数据库，会造成分布式数据不一致的情况，其根本原因是存储资源的独立性。

其实这种场景在实际生活中也会经常遇到。例如春晚演节目，只有每个节目都完美地演完，整个春晚才能说是完美，如果一个节目表演失败，然后中止后面的节目，并且将前面演过的节目进行回滚，这个显然是不现实的。为了保证春晚顺利完成，采取什么办法呢？彩排。就是先提前预演一遍，如果都没问题，在除夕夜正式表演。

映射到软件开发中，彩排可以类比为数据库预提交，先让每个服务（或者数据库）执行 SQL 语句，但是并不提交，如果每个服务（或者数据库）执行都没有问题，再让每个服务都提交（类似于除夕晚上统一上春晚）。如果一个服务有问题就回滚，那么自然就需要一个第三方协调者来统一协调各个服务（回想一下分布式锁也需要一个第三方。其实在分布式系统的解决方案中，基本都是通过引入第三方协调者来解决问题的）。

其实上面说的彩排和正式演出，可以将分布式事务的解决方案归为下面的两个阶段。

（1）准备阶段：第三方协调者向每个服务发送消息，让每个服务（或者数据库）执行本地事务，写本地的 redo 和 undo 日志，但不提交，达到一种"万事俱备，只欠东风"的状态。

（2）提交阶段：如果准备阶段一切正常，则进行真正的提交；如果准备阶段出现问题，则让每个服务（或者数据库）进行回滚。

10.3　X/Open 分布式事务模型

X/Open DTP 模型（X/Open distributed transaction processing reference model）是 X/Open 组织定义的一套分布式事务的标准，其定义了规范和 API 接口，由各个厂商进行具体的实现。这个思想也是 Java 平台一直遵循的。

X/Open DTP 定义了以下 3 个组件。

- ☑　AP（应用程序，application program）：可以理解为使用 DTP 的程序，也就是每个服务。
- ☑　TM（事务管理器，transaction manager）：负责协调和管理事务，提供给 AP 应用程序编程接口以及管理资源管理器。
- ☑　RM（资源管理器，resource manager）：这里可以理解为数据库，或者消息服务器管理系统，应用程序通过资源管理器对资源进行控制，资源必须实现 XA 定义的接口。

其中，AP 可以和 TM 以及 RM 通信，TM 和 RM 互相之间可以通信，X/Open DTP 模型定义了 XA 接口，TM 和 RM 通过 XA 接口进行双向通信，如图 10-3 所示。

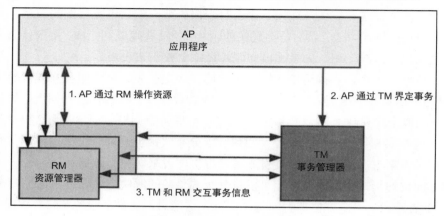

图 10-3　X/Open DTP 模型

例如，TM 通知 RM 提交事务或者回滚事务，RM 把提交结果通知给 TM。AP 和 RM 之间则通过 RM 提供的 Native API 进行资源控制，这个只定义了规范，各个厂商各自实现自己的资源控制。例如，Oracle 和 MySQL 都有自己的数据库驱动程序。

在生产中，使用 X/OpenDTP 的步骤如下。

（1）搭建 TM，将多个 RM 注册到 TM 中。

（2）AP 从 RM 中获取连接，如果 RM 是数据库，则获取 JDBC 连接，利用拿到的连接进行数据库的操作。

（3）AP 向 TM 发起一个全局事务，生成全局事务 ID，一般称作 XID。

（4）AP 通过获取的数据库连接操作 RM，这时，每次操作都会将 XID 传递给 RM。

（5）当 AP 结束全局事务时，TM 会通知各个 RM 全局事务结束，根据 RM 的反馈结果进行提交或回滚。

在这个模型中，TM 其实就是一个第三方，用来管理参与事务的多个 RM，TM 根据每个 RM 所执行的分支事务，进行提交或回滚。

TM 和 RM 直接的事务控制是基于 XA 协议来完成的，XA 协议是 X/Open 提出的分布式事务处理规范，也是分布式事务的工业标准，定义了交易中间件与数据库之间的接口规范（即接口函数），交易中间件用它来通知数据库事务的开始、结束以及提交、回滚等。XA 接口函数由数据库厂商提供，目前各大数据库厂商都实现了 XA 接口。

10.4 两阶段提交协议

本节将讲解两阶段提交协议的执行过程。

10.4.1 两阶段提交协议的过程

经过前面的分析知道，TM 管理多个 RM 的事务，其实涉及两个阶段：第一阶段是投票（准备），第二阶段是提交或者回滚。

其实两阶段提交类似于西式婚礼，婚礼两阶段的过程如下。

第一阶段，牧师对新郎说："你愿意娶这个女人吗？爱她、忠诚于她，无论她贫困、患病或者残疾，直至死亡。Do you（你愿意吗）？"

新郎："I do（我愿意）！"

牧师对新娘说："你愿意嫁给这个男人吗？爱他、忠诚于他，无论他贫困、患病或者残疾，直至死亡。Do you（你愿意吗）？"

新娘："I do（我愿意）！"

第二阶段，如果在第一阶段中新郎和新娘都反馈"I do"。

牧师对新郎和新娘说："现在请你们面向对方，握住对方的双手，作为妻子和丈夫向对方宣告誓言。"

如果第一阶段，新郎或者新娘有一人反馈 no。

牧师则发出回滚指令：这个婚不结了。

上面这个例子，其实就是两阶段的过程。首先 TM（牧师）会询问两个 RM（二位新人）是否能执行事务提交操作（愿意结婚）。如果两个 RM 能够执行事务的提交，首先执行事务操作，然后返回 yes，如果没有成功执行事务操作，就返回 no。

当 TM 接收到所有的 RM 的反馈之后，开始进入事务提交阶段。如果所有 RM 都返回 yes，那就发送 commit 请求，如果有一个人反馈 no，那就发出 rollback 请求。

在分布式事务中，两阶段提交过程如下。

1．准备/投票阶段

TM 给每个 RM 发送 prepare 消息，每个 RM 要么直接返回失败（如权限验证失败），要么在本地执行事务，写本地的 redo 和 undo 日志，但不提交。

这就是两阶段的第一阶段，分为以下三个步骤。

（1）TM 节点向所有 RM 节点询问是否可以执行提交操作，并开始等待各个 RM 节点的响应。

（2）RM 节点执行询问发起为止的所有事务操作，并将 undo 信息和 redo 信息写入日志（注意，若成功这里其实每个 RM 已经执行了事务操作）。

（3）各 RM 节点响应 TM 节点发起的询问。如果 RM 节点的事务操作实际执行成功，则它返回 yes 消息；如果 RM 节点的事务操作实际执行失败，则它返回 no 消息。

2．提交阶段

如果 TM 收到了各 RM 的失败消息或者超时，直接给每个 RM 发送回滚（rollback）消息；否则，发送提交（commit）消息；各个 RM 根据 TM 的指令执行提交或者回滚操作，释放所有事务处理过程中使用的锁资源（注意，必须在最后阶段释放锁资源）。

这就是两阶段的第二阶段，分为以下两种情况。

（1）当第一阶段从各 RM 获取的消息都为 yes 时，执行以下操作。

- TM 节点向所有 RM 节点发出"正式提交（commit）"的请求。
- RM 节点正式完成操作，并释放在整个事务期间占用的资源。
- RM 节点向 TM 节点发送"完成"消息。
- TM 节点收到所有 RM 节点反馈的"完成"消息后，即代表完成事务。

（2）如果任一 RM 节点在第一阶段返回的响应消息为 no，或者 TM 节点在第一阶段的询问超时之前，无法获取所有 RM 节点的响应消息时，则执行以下操作。

- TM 节点向所有 RM 节点发出"回滚操作（rollback）"的请求。
- RM 节点利用之前写入的 undo 信息执行回滚，并释放在整个事务期间内占用的资源。
- RM 节点向 TM 节点发送"回滚完成"消息。
- TM 节点收到所有 RM 节点反馈的"回滚完成"消息后，取消事务。

10.4.2　两阶段提交协议的缺点

两阶段提交能够提供原子性的操作，但两阶段提交还有以下 4 个缺点。

1．同步阻塞

执行过程中，所有 RM 都是事务阻塞型的。当 RM 占有公共资源时，其他第三方节点访问公共资源时，不得不处于阻塞状态。

2. 单点故障

由于 TM 的重要性，一旦 TM 发生故障。RM 会一直阻塞下去。尤其在第二阶段，TM 发生故障，那么所有的 RM 还都处于锁定事务资源的状态中，而无法继续完成事务操作（如果是 TM 宕机，可以重新选举一个 TM，但是无法解决因为 TM 宕机导致的 RM 处于阻塞状态的问题）。

3. 数据不一致

在两阶段提交的阶段二中，当 TM 向 RM 发送 commit 请求之后，发生了局部网络异常或者在发送 commit 请求过程中 TM 发生了故障，这会导致只有一部分 RM 接收到了 commit 请求。而在这部分 RM 接收到 commit 请求之后就会执行 commit 操作。但是其他部分未接收到 commit 请求的机器，则无法执行事务提交。于是整个分布式系统便出现了数据不一致性的现象。

4. 二阶段无法解决的问题

TM 在发出 commit 消息之后宕机，而唯一接收到这条消息的 RM 同时也宕机了。那么即使 TM 通过选举协议产生了新的 TM，这条事务的状态也是不确定的，因此没人知道事务是否已经被提交。

10.5 三阶段提交协议

由于两阶段提交协议存在诸如同步阻塞、单点故障、数据不一致等缺陷，所以，研究者在两阶段提交协议的基础上做了改进，提出了三阶段提交协议。

10.5.1 三阶段提交协议的过程

三阶段提交协议的流程如下。

1. CanCommit 阶段（询问阶段）

TM 向 RM 发送 commit 请求，RM 如果可以提交就返回 yes 响应，否则返回 no 响应。这阶段分为如下几个步骤。

（1）事务询问：TM 向 RM 发送 CanCommit 请求，询问是否可以执行事务提交操作。

然后开始等待 RM 的响应。

（2）响应反馈：RM 接到 CanCommit 请求之后，正常情况下，如果其自身认为可以顺利执行事务，则返回 yes 响应，并进入预备状态。否则反馈 no。

2．PreCommit 阶段（准备阶段）

TM 根据 RM 的反应情况，来决定是否可以进行事务的 PreCommit 操作。根据响应情况，有以下两种可能。

（1）假如 TM 从所有的 RM 获得的反馈都是 yes 响应，那么就会执行事务的预执行，即执行此阶段。

- ☑ 发送预提交请求：TM 向 RM 发送 PreCommit 请求，并进入 Prepared 阶段。
- ☑ 事务预提交：RM 接收到 PreCommit 请求后，会执行事务操作，并将 undo 和 redo 信息记录到事务日志中，但是不提交事务。
- ☑ 响应反馈：如果 RM 成功地执行了事务操作，则返回 ACK 响应，同时开始等待最终指令。

（2）假如有任何一个 RM 向 TM 发送了 no 响应，或者等待超时之后，TM 都没有接到 RM 的响应，那么就执行事务的中断。

- ☑ 发送中断请求：TM 向所有 RM 发送 abort 请求。
- ☑ 中断事务：RM 收到来自 TM 的 abort 请求之后（或超时之后）仍未收到 TM 的请求，执行事务的中断。

3．DoCommit（提交或回滚阶段）

该阶段进行真正的事务提交，也可以分为以下两种情况。

（1）TM 接收到所有 RM 发送的 ACK 响应都为 yes。

- ☑ 发送提交请求：TM 接收到 RM 发送的 ACK 响应，那么它将从预提交状态进入提交状态，并向所有 RM 发送 DoCommit 请求。
- ☑ 事务提交：RM 接收到 TM 的 DoCommit 请求之后，执行正式的事务提交，并在完成事务提交之后，释放所有事务资源。
- ☑ 响应反馈：事务提交完之后，向 TM 发送 ACK 响应。
- ☑ 完成事务：TM 接收到所有 RM 的 ACK 响应之后，完成事务。

（2）TM 没有接收到所有 RM 发送的 ACK 为 yes 的响应（可能是接收者发送的不是 ACK 响应，也可能响应超时）。

- ☑ 发送中断请求：TM 向所有 RM 发送 abort 请求。

- 事务回滚：RM 接收到 abort 请求之后，利用其在阶段二记录的 undo 信息来执行事务的回滚操作，并在完成回滚之后释放所有的事务资源。
- 反馈结果：RM 完成事务回滚之后，向 TM 发送确认消息。
- 中断事务：TM 接收到 RM 反馈的 ACK 消息之后，执行事务的中断。

在 DoCommit 阶段，如果 RM 无法及时接收到来自 TM 的 DoCommit 或者 rebort 请求时，会在等待超时之后，继续进行事务的提交。其实这个应该是基于概率来决定的，当进入第三阶段时，说明 RM 在第二阶段已经收到了 PreCommit 请求，那么 TM 产生 PreCommit 请求的前提条件是它在第二阶段开始之前，收到所有 RM 的 CanCommit 响应都是 yes。一旦 RM 收到了 PreCommit，意味着它知道大家其实都同意修改了。所以，一句话概括就是，当进入第三阶段时，由于网络超时等原因，虽然 RM 没有收到 commit 或者 abort 响应，但是它有理由相信，成功提交的概率很大。

10.5.2　两阶段提交协议和三阶段提交协议的区别

虽然三阶段提交协议在两阶段提交协议的基础上做了改进，但是也没有完全解决两阶段提交协议所存在的问题，三阶段提交协议和两阶段提交协议的具体区别如下。

（1）三阶段提交协议在两阶段提交协议的基础上增加了 CanCommit 阶段，用于询问所有的 RM 是否可以执行事务操作并且响应，它的好处是，在不占用资源的情况下，可以尽早发现无法执行的操作，从而中止后续的行为，减少资源的占用时间。

（2）TM 和 RM 都引入了超时机制。
- 当 TM 没有收到 RM 的响应时，TM 都会发出中止命令，让 RM 回滚。
- 当 RM 没有收到 TM 的命令时，如果 RM 处于 PreCommit 阶段，则进行回滚；如果 RM 处于 DoCommit 阶段，则进行提交。

10.6　CAP 定理和 BASE 理论

前面学习过的两阶段和三阶段提交协议，它们是 XA 协议解决数据一致性问题的基本方案，都保证了数据的强一致性，降低了可用性。实际上这里涉及分布式系统中的两个理论模型，即 CAP 定理和 BASE 理论。

10.6.1　CAP 定理

CAP 定理，又叫布鲁尔定理。指的是在一个分布式系统中，最多只能同时满足以下三项特性中的两项。

- ☑ 一致性（consistency），数据在多个副本中保持一致，可以理解成两个用户访问 A 和 B 两个系统，当 A 系统数据有变化时，及时同步给 B 系统，让两个用户看到的数据是一致的。
- ☑ 可用性（availability），系统对外提供服务必须一直处于可用状态，在任何故障下，客户端都能在合理时间内获得服务端非错误的响应。
- ☑ 分区容错性（partition tolerance），在分布式系统中遇到任何网络分区故障，系统仍然能对外提供服务。网络分区可以这样理解，在分布式系统中，不同的节点分布在不同的子网络中，有可能子网络中只有一个节点，在所有网络正常的情况下，由于某些原因导致这些子节点之间的网络出现故障，整个节点环境被切分成了不同的独立区域，这就是网络分区。

接下来详细分析 CAP，为什么只能满足三项中的两项，CAP 演示如图 10-4 所示。

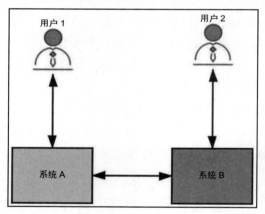

图 10-4　CAP 演示

用户 1 和用户 2 分别访问系统 A 和系统 B，系统 A 和系统 B 通过网络进行同步数据。理想情况是，用户 1 访问系统 A 对数据进行修改，将 data1 改成了 data2，同时用户 2 访问系统 B，拿到的是 data2 数据。

但是实际工作中，人们对分布式系统往往具有以下误解。

（1）网络相当可靠。

（2）延迟为零。

（3）传输带宽是无限的。

（4）网络相当安全。

（5）拓扑结构不会改变。

（6）必须要有一名管理员。

（7）传输成本为零。

（8）网络同质化。

只要有网络调用，网络总是不可靠的，下面我们来一一分析。

（1）当网络发生故障时，系统 A 和系统 B 无法进行数据同步，也就是不满足 P，同时两个系统依然可以访问，那么此时其实相当于是两个单机系统，就不是分布式系统了，若是分布式系统，P 必须满足。

（2）当 P 满足时，如果用户 1 通过系统 A 对数据进行了修改，将 data1 改成了 data2，也要让用户 2 通过系统 B 正确地拿到 data2，那么此时是满足 C，就必须等待网络将系统 A 和系统 B 的数据同步好，并且在同步期间，任何用户不能访问系统 B（让系统不可用），否则数据就不是一致的。此时满足的是 CP，即牺牲的是数据的可用性。

（3）当 P 满足时，如果用户 1 通过系统 A 对数据进行了修改，将 data1 改成了 data2，也要让系统 B 能继续提供服务，那么此时，只能接受系统 A 没有将 data2 同步给系统 B（牺牲了一致性）。此时满足的就是 AP，即牺牲的是数据的一致性。

在前面学过的注册中心 Eureka 就是满足的 AP，它并不保证 C。而 Zookeeper 是保证 CP，它不保证 A。在生产中，A 和 C 的选择没有标准的规定，它取决于自己的业务的。例如，12306 网站是满足 CP 的，因为买票业务必须满足数据的一致性。

10.6.2 BASE 理论

由于 CAP 中一致性 C 和可用性 A 无法兼得，eBay 的架构师提出了 BASE 理论，它是通过牺牲数据的强一致性，来获得可用性。BASE 理论具有如下 3 种特征。

- ☑ 基本可用（basically available）：分布式系统在出现不可预知故障时，允许损失部分可用性，保证核心功能的可用。
- ☑ 软状态（soft state）：软状态也称为弱状态，和硬状态相对，是指允许系统中的数据存在中间状态，并认为该中间状态的存在不会影响系统的整体可用性，即允许系统在不同节点的数据副本之间进行数据同步的过程存在延时。

☑ 最终一致性（eventually consistent）：最终一致性强调的是系统中所有的数据副本，在经过一段时间的同步后，最终能够达到一致的状态。因此，最终一致性的本质是需要系统保证最终数据能够达到一致，而不需要实时保证系统数据的强一致性。

BASE 理论并没有要求数据的强一致性，而是允许数据在一定的时间段内是不一致的，但在最终某个状态会达到一致。在生产环境中，很多公司会采用 BASE 理论来实现数据的一致，因为产品的可用性相比强一致性来说更加重要。例如在电商平台中，当用户对一个订单发起支付时，往往会调用第三方支付平台，如支付宝支付或者微信支付，调用第三方成功后，第三方并不能及时通知我方系统，在第三方没有通知我方系统的这段时间内，会将用户的订单状态显示为支付中，等第三方回调之后，再将状态改成已支付。虽然订单状态在短期内存在不一致，但是用户却获得了更好的产品体验。

10.7 TCC 分布式事务解决方案

前面介绍了分布式事务的理论模型，即两阶段提交协议和三阶段提交协议。本节来学习在生产环境中实际采用的一些解决方案，先从 TCC 方案开始。

10.7.1 TCC 方案

TCC 方案（try-confirm-cancel）是一种常用的分布式事务解决方案，它将一个事务拆分成 3 个步骤。

（1）T（try）：业务检查阶段，这阶段主要进行业务校验和检查或者资源预留；也可能是直接进行业务操作。

（2）C（confirm）：业务确认阶段，这阶段对 try 阶段校验过的业务或者预留的资源进行确认。

（3）C（cancel）：业务回滚阶段，这阶段和上面的 C（confirm）是互斥的，用于释放 try 阶段预留的资源或者业务。

其实 TCC 是一种两阶段提交方案，第一阶段就是 try，进行业务校验或者资源预留，第二阶段就是根据第一阶段的结果，来决定是进行 confirm 还是 cancel。TCC 方案的整体流程如图 10-5 所示。

在 confirm 和 cancel 阶段，如果执行出错，需要采用重试机制或者人工介入。

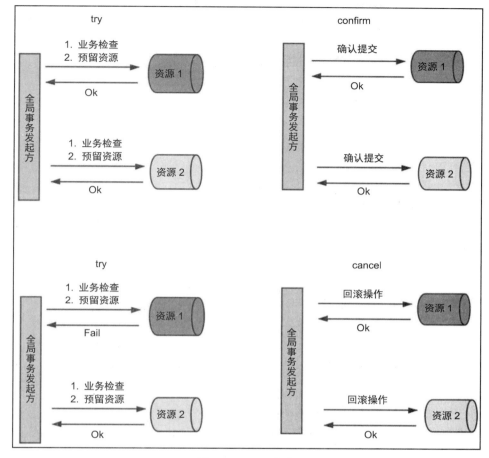

图 10-5　TCC 方案

在实际工作中，对 TCC 的使用可以灵活变通，一般采用以下两种方式。

1．加预留冻结字段

例如，在转账业务中，张三向李四转账 100 元，张三的余额表会多增加一个冻结金额字段，当转账业务发起时，try 阶段执行下面的操作。

（1）张三的余额减 100 元，张三的冻结金额字段加 100 元。

（2）李四的冻结金额字段加 100 元。

如果 try 阶段都返回正常，则执行 confirm。confirm 阶段执行下面的操作。

（1）张三的冻结金额字段减 100 元。

（2）李四的冻结金额字段减 100 元，李四的余额加 100 元。

如果 try 阶段返回异常，则执行 cancel。cancel 阶段执行下面的操作。

（1）张三的余额加 100，张三的冻结金额字段减 100。

（2）李四的冻结金额字段减 100。

2. 直接操作数据

下面还以转账业务为例。

try 阶段执行下面的操作。

（1）张三的余额减 100。

（2）李四的余额加 100。

如果 try 阶段都返回正常，则执行 confirm。confirm 阶段执行下面的操作。

（1）张三执行空。

（2）李四执行空。

如果 try 阶段返回异常，则执行 cancel。cancel 阶段执行下面的操作。

（1）张三的余额加 100。

（2）李四的余额减 100。

10.7.2　TCC 方案的异常处理

TCC 方案还需要处理以下 3 种异常。

1. 空回滚

在没有调用 TCC 资源 try 方法的情况下，调用了第二阶段的 cancel 方法。例如上面的例子中，当 try 请求由于网络延迟或故障等原因没有执行，结果返回了异常，那么此时 cancel 就不能正常执行，因为 try 没有对数据进行修改，如果 cancel 进行了对数据的修改，那就会导致数据不一致。

解决问题的关键是要识别出这个操作是一个空回滚。思路很简单，需要知道 try 阶段是否执行，如果执行了，那就是正常回滚；如果没执行，那就是空回滚。前面已经说过 TM 在发起全局事务时生成全局事务记录，全局事务 ID 贯穿整个分布式事务调用链条。再额外增加一张分支事务记录表，其中有全局事务 ID 和分支事务 ID，第一阶段 try 方法里会插入一条记录表示 try 阶段执行了。cancel 接口里读取该记录，如果该记录存在，则正常回滚；如果该记录不存在，则是空回滚。

2. 幂等

通过前面介绍已经了解到，为了保证 TCC 二阶段提交，会采用重试机制来保证请求的执行，这样就要求 TCC 的二阶段 confirm 和 cancel 接口保证幂等，这样不会重复使用或者释放资源。如果幂等控制没有做好，很有可能会导致数据不一致等严重问题。

解决思路是，在上述的分支事务记录中增加执行状态，每次执行前都查询该状态。

3. 悬挂

悬挂就是对于一个分布式事务来说，其二阶段 cancel 接口比 try 接口先执行。

原因是在调用分支事务 try 时，由于网络发生拥堵，造成了超时，TM 就会通知 RM 回滚该分布式事务，可能回滚完成后，try 请求才到达参与者真正执行，一个 try 方法预留的业务资源，只有该分布式事务才能使用，这样该分布式事务第一阶段预留的业务资源，就再也没有人能够处理了，对于这种情况就称为悬挂，即业务资源预留后无法继续处理。

解决思路是，如果二阶段执行完成，那一阶段就不能再继续执行。在执行一阶段事务时判断在该全局事务下，分支事务记录表中是否已经有二阶段的事务记录，如果有，则不执行 try。

从上面的 3 个异常情况来看，需要在 TCC 解决方案中增加一张分支事务记录表，如表 10-1 所示。

表 10-1 分支事务记录表

字段	描述
tx_id	全局事务 id
branch_id	分支 id
tx_status	事务状态： 1：try 执行后赋值为事务初始化 2：confirm 执行后赋值为已提交 3：cancel 执行后赋值为已回滚

对于空回滚问题，可以判断 try 是否执行，如果 try 未执行，则为空回滚。

对于幂等问题，每次执行前判断数据库的状态，如果状态已经变更，则不再执行。

对于悬挂问题，分以下两步进行处理。

（1）执行 cancel 之前判断 try 有没有执行，如果没执行，则 cancel 执行空方法，并且插入分支事务记录表，记录 cancel 已经执行过了。

（2）执行 try 之前判断是否有 cancel 执行过，如果执行过，则 try 执行空方法。

10.8 可靠消息最终一致性方案

下面根据前面的 BASE 理论学习可靠消息最终一致性方案。它是保证事务最终一致性的一种方案，允许数据在业务中出现短暂的不一致状态。

可靠消息最终一致性方案是指当事务发起方（事务参与者，也就是消息的发送者）执行完本地事务后，同时发出一条消息，事务消费方（事务参与者，也就是消息的消费者）一定能够接收消息并可以成功地处理自己的事务，如图 10-6 所示。

图 10-6　可靠消息服务最终一致性方案

需要注意以下两点。
- ☑ 可靠消息：事务发起方一定得把消息传递到事务消费方。
- ☑ 最终一致性：最终事务发起方的业务处理和事务消费方的业务处理得以完成，并达成最终一致。

10.8.1 可靠消息最终一致性问题分析

从图 10-6 可以看出，事务发起方将消息发送给消息中间件，事务消费方从消息中间件接收消息，事务发起方和消息中间件之间、事务消费方和消息中间件之间都有网络通信，由于网络通信的不确定性，这将会导致数据的问题。下面针对导致的问题来分别进行解决。

1. 事务发起方本地事务和消息发送之间的原子性问题

此问题是本地事务执行成功，消息必须发出去，否则丢弃消息，即本地事务执行和消息的发送，要么都成功，要么都失败。

接下来参考以下伪代码。

```
begin transaction;
发送消息;
```

```
操作数据库;
commit transaction;
```

此时,如果发送消息成功,数据库操作失败,则无法保证原子性。下面调换发送消息和操作数据库的顺序。

```
begin transaction;
操作数据库;
发送消息;
commit transaction;
```

此时,如果操作数据库出错时,回滚,不影响数据;如果发送消息出错时,也回滚,不影响数据。这样看来似乎可以保证原子性,但是会有一种情况,即发送消息响应超时,导致数据库回滚,但是消息已经发送成功了,这时原子性则是无法保证的,此时就需要人工补偿了。

2. 事务消费方和消息消费的原子性问题

此时要保证事务消费方必须能接收到消息,如果由于程序故障导致事务消费方重启,那么需要消息中间件要有消息重发机制;由于网络延时的存在,当事务消费方消费消息成功,没有向消息中间件响应时,而消息中间件由于重发机制再次投递消息时,就会导致消息重复消费的问题。此时,在消费方要有幂等性解决方案。

10.8.2 本地消息事件表方案

通过在系统中增加一张本地消息事件表 tbl_local_event 可以解决 10.8.1 节出现的问题。本地消息事件表的结构如表 10-2 所示。

表 10-2 本地消息事件表结构

字 段 名	描 述
id	事件唯一标识。每个事件在创建时都会生成一个全局唯一 ID,如 UUID
event_status	事件状态。枚举类型。现在只有两个状态:待发布(NEW)和已发布(PUBLISHED)
payload	这里会将事件内容转成 JSON 存到这个字段中
event_type	事件类型。枚举类型。每个事件都会有一个类型,如业务中的创建用户 USER_CREATED 就是一个事件类型

本地消息事件表流程如图 10-7 所示。

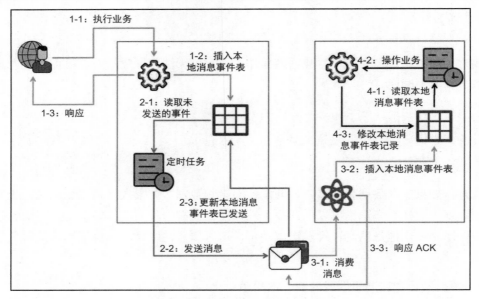

图 10-7 本地消息事件表流程

本地消息事件表流程详解如下。

（1）执行业务，插入本地消息事件表。

- ☑ 执行业务。
- ☑ 插入本地消息事件表。此时事件的状态是新建。
- ☑ 向客户端响应。

（2）读取本地消息事件表，发送消息。

- ☑ 通过定时任务去读取本地消息事件表中新建状态的记录。
- ☑ 将记录发送给消息队列。
- ☑ 更新事件状态为已发送。

（3）接收消息，插入本地消息事件表。

- ☑ 从消息队列消费消息。
- ☑ 插入本地消息事件表，此时事件状态为已接收。
- ☑ 响应消息队列，发送 ACK。

（4）读取本地消息事件表，执行业务。

- ☑ 读取本地消息事件表中已接收状态的记录。
- ☑ 执行业务操作。
- ☑ 修改事件状态为已完成。

以上流程中的每一个步骤都可以通过本地事务来保证，通过本地事务结合消息队列，解决了本地业务和发送消息的原子性问题，以及消息消费和执行业务的原子性问题。还可以通过事件表的主键冲突，解决重复消费消息的幂等性的问题。

10.8.3　RocketMQ 事务消息方案

此方案需要借助一个消息中间件——RocketMQ，它是阿里巴巴的一个开源消息中间件，阿里集团内部的消息都运行在 RocketMQ 之上，可见它的性能之强。它在 4.3 之后的版本支持了事务消息，为解决分布式事务提供了便利。

RocketMQ 的事务消息主要是为了解决事务生产方执行业务和消息发送的原子性问题。事务消息流程如图 10-8 所示。

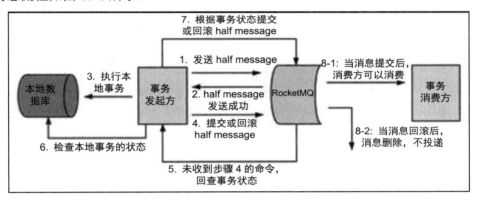

图 10-8　事务消息流程

以下是事务消息的几种状态。

- ☑ TransactionStatus.CommitTransaction：提交状态，它允许消费方消费此消息。
- ☑ TransactionStatus.RollbackTransaction：回滚状态，它代表该消息将被删除，不允许被消费方消费。
- ☑ TransactionStatus.Unknown：中间状态，它代表需要检查消息队列来确定状态。

事务消息的具体流程如下。

（1）发送 half message。在执行本地业务之前，先向消息队列发送一条事务消息，此时叫作 half message，此时消息被标记为 Prepared 预备状态，此时的消息是无法被消费方消费的，需要生产者对消息进行二次确认后，消费方才能去消费。

（2）消息队列回应 half message 发送成功。

（3）当事务发起方收到消息队列的发送成功响应之后，开始执行本地业务。

（4）如果本地事务执行成功，则向消息队列发送 half message 的确认，这样事务消费方就可以消费消息了。

（5）如果本地事务执行失败，则向消息队列发送 half message 的回滚，删除 half message，事务消费方就无法消费消息。

（6）回查机制。当第（4）步（无论是提交还是回滚）由于网络闪断，生产者应用重启等原因，导致生产者无法对消息队列中的 half message 进行二次确认（即上面的步骤（4），发送提交或者回滚消息）时，消息队列中的 half message 就不知道应该怎么办了。此时消息队列会定时扫描长期处于 half message 的消息，并发起一个回查机制来确认此时的 half message 应该是提交还是回滚。此时，消息队列主动询问生产者该消息的最终状态（提交还是回滚），即为消息的回查机制。

10.9　RocketMQ 安装部署

RocketMQ 的安装部署步骤如下。

（1）下载 RocketMQ，打开下载地址 http://rocketmq.apache.org/dowloading/releases/，RocketMQ 的下载界面如图 10-9 所示。

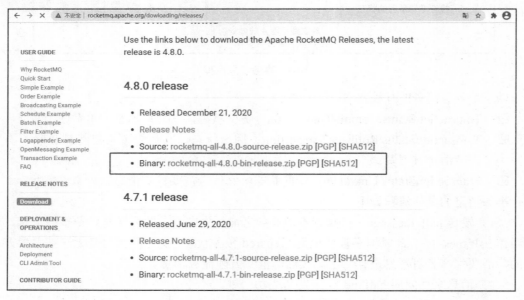

图 10-9　RocketMQ 下载界面

这里下载的是 4.8.0 版本。下载后得到一个 rocketmq-all-4.8.0-bin-release.zip 文件。

（2）解压安装包 rocketmq-all-4.8.0-bin-release.zip，得到目录 rocketmq-all-4.8.0-bin-release。

（3）配置环境变量，在系统变量中添加 ROCKETMQ_HOME，它的值为 E:\rocketmq\rocketmq-all-4.8.0-bin-release，如图 10-10 所示。

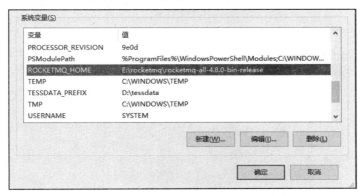

图 10-10 配置环境变量

（4）启动 NameServer，进入目录 E:\rocketmq\rocketmq-all-4.8.0-bin-release\bin，执行以下命令。

```
start mqnamesrv.cmd
```

命令执行成功后会弹出提示框，此框勿关闭。此时 NameServer 的地址为 127.0.0.1:9876。

（5）启动 Broker，进入目录 E:\rocketmq\rocketmq-all-4.8.0-bin-release\bin，执行如下命令。

```
start mqbroker.cmd -n 127.0.0.1:9876 autoCreateTopicEnable=true
```

其中的参数说明如下。
- ☑ -n：配置 NameServer 地址。
- ☑ autoCreateTopicEnable：是否自动创建 topic。

（6）下载可视化组件。

```
$ git clone https://github.com/apache/rocketmq-externals.git
Cloning into 'rocketmq-externals'...
remote: Enumerating objects: 33, done.
remote: Counting objects: 100% (33/33), done.
remote: Compressing objects: 100% (20/20), done.
```

```
remote: Total 19009 (delta 4), reused 14 (delta 2), pack-reused 18976
Receiving objects: 100% (19009/19009), 33.27 MiB | 51.00 KiB/s, done.
Resolving deltas: 100% (7416/7416), done.
Updating files: 100% (6192/6192), done.
```

下载后得到一个文件夹 rocketmq-externals。

（7）配置 application.properties。

进入目录 E:\rocketmq\rocketmq-externals\rocketmq-console\src\main\resources，修改配置文件，代码如下所示。

```
server.contextPath=
server.port=8080

### SSL setting
#server.ssl.key-store=classpath:rmqcngkeystore.jks
#server.ssl.key-store-password=rocketmq
#server.ssl.keyStoreType=PKCS12
#server.ssl.keyAlias=rmqcngkey

#spring.application.index=true
spring.application.name=rocketmq-console
spring.http.encoding.charset=UTF-8
spring.http.encoding.enabled=true
spring.http.encoding.force=true
logging.config=classpath:logback.xml
#if this value is empty,use env value rocketmq.config.namesrvAddr
NAMESRV_ADDR | now, you can set it in ops page.default localhost:9876
rocketmq.config.namesrvAddr=localhost:9876
#if you use rocketmq version < 3.5.8, rocketmq.config.isVIPChannel should
be false.default true
rocketmq.config.isVIPChannel=
#rocketmq-console's data path:dashboard/monitor
rocketmq.config.dataPath=/tmp/rocketmq-console/data
#set it false if you don't want use dashboard.default true
rocketmq.config.enableDashBoardCollect=true
#set the message track trace topic if you don't want use the default one
rocketmq.config.msgTrackTopicName=
```

```
rocketmq.config.ticketKey=ticket

#Must create userInfo file: ${rocketmq.config.dataPath}/users.properties if
the login is required
rocketmq.config.loginRequired=false
```

（8）打包可视化组件。

进入目录 E:\rocketmq\rocketmq-externals\rocketmq-console，执行如下命令。

```
mvn clean package -Dmaven.test.skip=true
```

命令执行完成后，在目录 E:\rocketmq\rocketmq-externals\rocketmq-console\target 中得到文件 rocketmq-console-ng-1.0.1.jar。

（9）启动可视化组件。

在步骤（8）的目录下，执行如下命令。

```
java -jar rocketmq-console-ng-1.0.1.jar
```

访问 http://localhost:8080，可视化组件界面如图 10-11 所示，此时的 8080 端口是在步骤（7）中配置的。

图 10-11　可视化组件界面

10.10　RocketMQ 事务消息实战

此案例通过一个生产者和一个消费者,来模拟事务消息解决方案。需要准备两个 Web 项目,即 producer 和 consumer。通过两个项目共同完成用户下单、创建订单、增加积分等任务。具体实现如下。

(1) producer 项目的需求是创建基础订单,并向消息队列发送消息。

- ☑ 在基础订单表 order_base 中插入一条数据,字段分别为 id(id 自增)和 order_no(业务订单号)。
- ☑ 在事务日志表 transaction_log 中插入一条数据,字段分别为 id(事务 id 号)、business (order_base 表中的业务订单号 order_no)和 foreign_key(此字段管理 order_base 表中的 id)。
- ☑ 向 RocketMQ 发送一条消息,topic 为 order-create,内容为订单的业务号。

上述 3 个操作是原子性的,即向 order_base 和 transaction_log 两张表中插入数据和向消息队列发送消息是原子操作。

(2) consumer 项目的需求是从消息队列订阅消息,然后给用户增加积分。

- ☑ 订阅消息队列中的 topic 为 order-create 的消息。
- ☑ 向 tbl_points 表中插入一条记录。
- ☑ 向消息队列回复 ACK 消息。

10.10.1　生产者 producer

producer 项目的目录结构如图 10-12 所示。

具体的实现步骤如下。

(1) 数据库准备。

创建数据库 mq-producer,在数据库中创建下面两张表。

第一张表 order_base,创建表的具体语句如下。

```
CREATE TABLE `order_base` (
  `id` int(16) NOT NULL AUTO_INCREMENT,
  `order_no` varchar(32) DEFAULT NULL,
```

```
  PRIMARY KEY (`id`)
) ENGINE=InnoDB AUTO_INCREMENT=35 DEFAULT CHARSET=utf8;
```

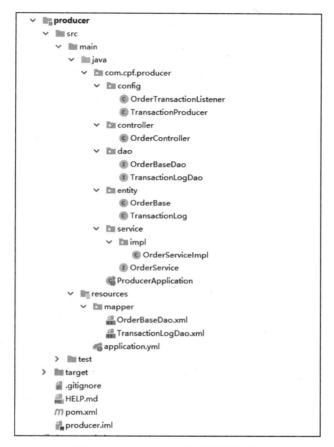

图 10-12　producer 项目的目录结构

第二张表 transaction_log，创建表的具体语句如下。

```
CREATE TABLE `transaction_log` (
  `id` varchar(32) CHARACTER SET utf8mb4 COLLATE utf8mb4_bin NOT NULL COMMENT '事务 ID',
  `business` varchar(32) CHARACTER SET utf8mb4 COLLATE utf8mb4_bin NOT NULL COMMENT '业务标识',
  `foreign_key` varchar(32) CHARACTER SET utf8mb4 COLLATE utf8mb4_bin NOT NULL COMMENT '对应业务表中的主键',
```

```
  PRIMARY KEY (`id`)
) ENGINE=InnoDB DEFAULT CHARSET=utf8mb4 COLLATE=utf8mb4_bin;
```

（2）修改项目的 pom 文件。

```xml
<?xml version="1.0" encoding="UTF-8"?>
<project xmlns="http://maven.apache.org/POM/4.0.0" xmlns:xsi="http://www.w3.org/2001/XMLSchema-instance"
    xsi:schemaLocation="http://maven.apache.org/POM/4.0.0 https://maven.apache.org/xsd/maven-4.0.0.xsd">
    <modelVersion>4.0.0</modelVersion>
    <parent>
        <groupId>org.springframework.boot</groupId>
        <artifactId>spring-boot-starter-parent</artifactId>
        <version>2.3.3.RELEASE</version>
        <relativePath/> <!-- lookup parent from repository -->
    </parent>
    <groupId>com.cpf.producer</groupId>
    <artifactId>producer</artifactId>
    <version>0.0.1-SNAPSHOT</version>
    <name>producer</name>
    <description>Demo project for Spring Boot</description>

    <properties>
        <java.version>1.8</java.version>
    </properties>

    <dependencies>
        <!--rocketmq-->
        <dependency>
            <groupId>org.apache.rocketmq</groupId>
            <artifactId>rocketmq-spring-boot-starter</artifactId>
            <version>2.0.2</version>
        </dependency>

        <!-- json-lib -->
        <dependency>
            <groupId>net.sf.json-lib</groupId>
```

```xml
        <artifactId>json-lib</artifactId>
        <version>2.4</version>
        <classifier>jdk15</classifier>
</dependency>


<dependency>
    <groupId>org.springframework.boot</groupId>
    <artifactId>spring-boot-starter-web</artifactId>
</dependency>

<dependency>
    <groupId>org.springframework.boot</groupId>
    <artifactId>spring-boot-starter-test</artifactId>
    <scope>test</scope>
    <exclusions>
        <exclusion>
            <groupId>org.junit.vintage</groupId>
            <artifactId>junit-vintage-engine</artifactId>
        </exclusion>
    </exclusions>
</dependency>

<dependency>
    <groupId>org.springframework</groupId>
    <artifactId>spring-tx</artifactId>
    <version>5.2.8.RELEASE</version>
</dependency>

<dependency>
    <groupId>org.springframework.boot</groupId>
    <artifactId>spring-boot-starter-web</artifactId>
</dependency>

<!-- MyBatis 相关依赖 -->
<dependency>
    <groupId>org.mybatis.spring.boot</groupId>
```

```xml
    <artifactId>mybatis-spring-boot-starter</artifactId>
    <version>2.0.0</version>
</dependency>

<!-- MySQL 驱动 -->
<dependency>
    <groupId>mysql</groupId>
    <artifactId>mysql-connector-java</artifactId>
</dependency>

<!-- 阿里巴巴数据库连接池 -->
<dependency>
    <groupId>com.alibaba</groupId>
    <artifactId>druid</artifactId>
    <version>1.1.12</version>
</dependency>

<dependency>
    <groupId>org.projectlombok</groupId>
    <artifactId>lombok</artifactId>
    <optional>true</optional>
</dependency>


<dependency>
    <groupId>org.springframework.boot</groupId>
    <artifactId>spring-boot-starter</artifactId>
</dependency>

<dependency>
    <groupId>org.springframework.boot</groupId>
    <artifactId>spring-boot-starter-test</artifactId>
    <scope>test</scope>
    <exclusions>
        <exclusion>
            <groupId>org.junit.vintage</groupId>
            <artifactId>junit-vintage-engine</artifactId>
```

```xml
                </exclusion>
            </exclusions>
        </dependency>
    </dependencies>

    <build>
        <plugins>
            <plugin>
                <groupId>org.springframework.boot</groupId>
                <artifactId>spring-boot-maven-plugin</artifactId>
            </plugin>
        </plugins>
    </build>
</project>
```

(3) 修改项目的 application.yml 文件。

```yaml
server:
  port: 8081

spring:
  application:
    name: producer
  datasource:
    type: com.alibaba.druid.pool.DruidDataSource
    driver-class-name: com.mysql.cj.jdbc.Driver
    url: jdbc:mysql://localhost:3306/mq-producer?characterEncoding=UTF-8&serverTimezone=Asia/Shanghai
    username: root
    password: root
    dbcp2:
      initial-size: 5
      min-idle: 5
      max-total: 5
      max-wait-millis: 200
      validation-query: SELECT 1
      test-while-idle: true
```

```yaml
      test-on-borrow: false
      test-on-return: false

mybatis:
  mapper-locations:
  - classpath:mapper/*.xml
```

（4）编写 mapper 文件。

在 resources 目录下创建目录 mapper。编写数据库访问的 mapper 文件。

表 order_base 的操作文件 OrderBaseDao.xml 如下所示。

```xml
<?xml version="1.0" encoding="UTF-8"?>
<!DOCTYPE mapper PUBLIC "-//mybatis.org//DTD Mapper 3.0//EN" "http://mybatis.org/dtd/mybatis-3-mapper.dtd">
<mapper namespace="com.cpf.producer.dao.OrderBaseDao">
  <resultMap id="BaseResultMap" type="com.cpf.producer.entity.OrderBase">
      <id column="id" jdbcType="INTEGER" property="id" />
      <result column="order_no" jdbcType="VARCHAR" property="orderNo" />
  </resultMap>
  <sql id="Base_Column_List">
      id, order_no
  </sql>
 <select id="selectByPrimaryKey" parameterType="java.lang.Integer" resultMap="BaseResultMap">
      select
      <include refid="Base_Column_List" />
      from order_base
      where id = #{id,jdbcType=INTEGER}
  </select>
  <delete id="deleteByPrimaryKey" parameterType="java.lang.Integer">
      delete from order_base where id = #{id,jdbcType=INTEGER}
  </delete>
  <insert id="insert" keyColumn="id" keyProperty="id" parameterType="com.cpf.producer.entity.OrderBase" useGeneratedKeys="true">
      insert into order_base (order_no) values (#{orderNo,jdbcType=VARCHAR})
  </insert>
  <insert id="insertSelective" keyColumn="id" keyProperty="id"
```

```xml
    parameterType="com.cpf.producer.entity.OrderBase" useGeneratedKeys="true">
        insert into order_base
        <trim prefix="(" suffix=")" suffixOverrides=",">
            <if test="orderNo != null">
                order_no,
            </if>
        </trim>
        <trim prefix="values (" suffix=")" suffixOverrides=",">
            <if test="orderNo != null">
                #{orderNo,jdbcType=VARCHAR},
            </if>
        </trim>
    </insert>
    <update id="updateByPrimaryKeySelective" parameterType="com.cpf.producer.entity.OrderBase">
        update order_base
        <set>
            <if test="orderNo != null">
                order_no = #{orderNo,jdbcType=VARCHAR},
            </if>
        </set>
        where id = #{id,jdbcType=INTEGER}
    </update>
    <update id="updateByPrimaryKey" parameterType="com.cpf.producer.entity.OrderBase">
        update order_base
        set order_no = #{orderNo,jdbcType=VARCHAR}
        where id = #{id,jdbcType=INTEGER}
    </update>
</mapper>
```

表 transaction_log 的操作文件 TransactionLogDao.xml 如下所示。

```xml
<?xml version="1.0" encoding="UTF-8"?>
<!DOCTYPE mapper PUBLIC "-//mybatis.org//DTD Mapper 3.0//EN" "http://mybatis.org/dtd/mybatis-3-mapper.dtd">
<mapper namespace="com.cpf.producer.dao.TransactionLogDao">
    <resultMap id="BaseResultMap" type="com.cpf.producer.entity.
```

```xml
TransactionLog">
    <id column="id" jdbcType="VARCHAR" property="id" />
    <result column="business" jdbcType="VARCHAR" property="business" />
    <result column="foreign_key" jdbcType="VARCHAR" property="foreignKey" />
  </resultMap>
  <sql id="Base_Column_List">
    id, business, foreign_key
  </sql>
  <select id="selectByPrimaryKey" parameterType="java.lang.String" resultMap="BaseResultMap">
    select
    <include refid="Base_Column_List" />
    from transaction_log
    where id = #{id,jdbcType=VARCHAR}
  </select>
  <delete id="deleteByPrimaryKey" parameterType="java.lang.String">
    delete from transaction_log
    where id = #{id,jdbcType=VARCHAR}
  </delete>
  <insert id="insert" keyColumn="id" keyProperty="id" parameterType="com.cpf.producer.entity.TransactionLog" useGeneratedKeys="true">
    insert into transaction_log (id,business, foreign_key)
    values (#{id,jdbcType=VARCHAR},#{business,jdbcType=VARCHAR},
    #{foreignKey,jdbcType=VARCHAR})
  </insert>
  <insert id="insertSelective" keyColumn="id" keyProperty="id" parameterType="com.cpf.producer.entity.TransactionLog" useGeneratedKeys="true">
    insert into transaction_log
    <trim prefix="(" suffix=")" suffixOverrides=",">
      <if test="business != null">
        business,
      </if>
      <if test="foreignKey != null">
        foreign_key,
      </if>
    </trim>
    <trim prefix="values (" suffix=")" suffixOverrides=",">
```

```xml
        <if test="business != null">
          #{business,jdbcType=VARCHAR},
        </if>
        <if test="foreignKey != null">
          #{foreignKey,jdbcType=VARCHAR},
        </if>
      </trim>
  </insert>
  <update id="updateByPrimaryKeySelective" parameterType="com.cpf.producer.entity.TransactionLog">
    update transaction_log
    <set>
      <if test="business != null">
        business = #{business,jdbcType=VARCHAR},
      </if>
      <if test="foreignKey != null">
        foreign_key = #{foreignKey,jdbcType=VARCHAR},
      </if>
    </set>
    where id = #{id,jdbcType=VARCHAR}
  </update>
  <update id="updateByPrimaryKey" parameterType="com.cpf.producer.entity.TransactionLog">
    update transaction_log
    set business = #{business,jdbcType=VARCHAR},
      foreign_key = #{foreignKey,jdbcType=VARCHAR}
    where id = #{id,jdbcType=VARCHAR}
  </update>

  <select id="selectCount" parameterType="java.lang.String" resultType="java.lang.Integer">
    select
    count(*)
    from transaction_log
    where id = #{id,jdbcType=VARCHAR}
  </select>
</mapper>
```

（5）编写实体类。

在com.cpf.producer.entity包下创建类OrderBase.java，代码如下。

```java
package com.cpf.producer.entity;

import java.io.Serializable;
import lombok.Data;

/**
 * order_base
 * @author 晁鹏飞
 */
@Data
public class OrderBase implements Serializable {
    private Integer id;

    private String orderNo;

}
```

创建类TransactionLog.java，代码如下。

```java
package com.cpf.producer.entity;

import java.io.Serializable;
import lombok.Data;

/**
 * transaction_log
 * @author 晁鹏飞
 */
@Data
public class TransactionLog implements Serializable {
    /**
     * 事务id
     */
    private String id;
```

```java
/**
 * 订单业务号
 */
private String business;

/**
 * 对应业务表中的主键
 */
private String foreignKey;

}
```

（6）编写对应的 DAO 操作类。

在 com.cpf.producer.dao 包下，创建类 OrderBaseDao.java，代码如下。

```java
package com.cpf.producer.dao;

import com.cpf.producer.entity.OrderBase;
import org.apache.ibatis.annotations.Mapper;

/**
 * OrderBase 数据操作层
 * @author 晁鹏飞
 */
@Mapper
public interface OrderBaseDao {
    int deleteByPrimaryKey(Integer id);

    int insert(OrderBase record);

    int insertSelective(OrderBase record);

    OrderBase selectByPrimaryKey(Integer id);

    int updateByPrimaryKeySelective(OrderBase record);

    int updateByPrimaryKey(OrderBase record);
}
```

创建类 TransactionLogDao.java，代码如下。

```java
package com.cpf.producer.dao;

import com.cpf.producer.entity.TransactionLog;
import org.apache.ibatis.annotations.Mapper;

/**
 * TransactionLog 数据操作层
 * @author 晁鹏飞
 */
@Mapper
public interface TransactionLogDao {

    int deleteByPrimaryKey(String id);

    int insert(TransactionLog record);

    int insertSelective(TransactionLog record);

    TransactionLog selectByPrimaryKey(String id);

    int updateByPrimaryKeySelective(TransactionLog record);

    int updateByPrimaryKey(TransactionLog record);

    int selectCount(String id );
}
```

（7）编写 Service 层。

创建包 com.cpf.producer.service，定义接口 OrderService.java，代码如下。

```java
package com.cpf.producer.service;

import com.cpf.producer.entity.OrderBase;
import org.apache.rocketmq.client.exception.MQClientException;

import java.lang.reflect.InvocationTargetException;
```

```java
/**
 * 订单操作接口
 * @author 晁鹏飞
 */
public interface OrderService {
    /**
     * 创建订单，实际接口
     * @param order
     * @param transactionId
     * @throws InvocationTargetException
     * @throws IllegalAccessException
     */
    public void createOrder(OrderBase order, String transactionId)
throws InvocationTargetException, IllegalAccessException;

    /**
     * 创建订单，发送消息接口
     * @param order
     * @throws MQClientException
     */
    public void createOrder(OrderBase order) throws MQClientException;
}
```

在包 com.cpf.producer.service 下创建子包 impl，定义接口 OrderService 的实现类 OrderServiceImpl.java，代码如下。

```java
package com.cpf.producer.service.impl;

import com.alibaba.fastjson.JSON;
import com.cpf.producer.config.TransactionProducer;
import com.cpf.producer.dao.TransactionLogDao;
import com.cpf.producer.entity.TransactionLog;
import com.cpf.producer.service.OrderService;
import com.cpf.producer.dao.OrderBaseDao;
import com.cpf.producer.dao.TransactionLogDao;
import com.cpf.producer.entity.OrderBase;
import com.cpf.producer.entity.TransactionLog;
```

```java
import com.cpf.producer.service.OrderService;
import lombok.extern.slf4j.Slf4j;
import org.apache.rocketmq.client.exception.MQClientException;
import org.springframework.beans.factory.annotation.Autowired;
import org.springframework.stereotype.Service;
import org.springframework.transaction.annotation.Transactional;

import java.lang.reflect.InvocationTargetException;

/**
 * OrderService 的实现类
 * @author 晁鹏飞
 */
@Service
@Slf4j
public class OrderServiceImpl implements OrderService {
    @Autowired
    OrderBaseDao orderMapper;
    @Autowired
    TransactionLogDao transactionLogMapper;
    @Autowired
    TransactionProducer producer;

    /**
     * 执行本地事务时调用,将订单数据和事务日志写入本地数据库
     * @param order
     * @param transactionId
     * @throws InvocationTargetException
     * @throws IllegalAccessException
     */
    @Transactional
    @Override
    public void createOrder(OrderBase order, String transactionId)
throws InvocationTargetException, IllegalAccessException {
        System.out.println(transactionId);

        // 1.创建订单
```

```
            orderMapper.insert(order);

            // 2.写入事务日志
            TransactionLog log = new TransactionLog();
            log.setId(transactionId);
            log.setBusiness(order.getOrderNo());
            log.setForeignKey(String.valueOf(order.getId()));
            transactionLogMapper.insert(log);

        }

        /**
         * controller 调用,只用于向 RocketMQ 发送事务消息
         * @param order
         * @throws MQClientException
         */
        @Override
        public void createOrder(OrderBase order) throws MQClientException {
            // 向RocketMQ 发送消息, topic 为 order-create
            producer.send(JSON.toJSONString(order),"order-create");
        }
    }
```

(8) 创建事务监听器。

创建包 com.cpf.producer.config,在此包下创建事务监听器 OrderTransactionListener.java,代码如下。

```
package com.cpf.producer.config;

import com.alibaba.fastjson.JSONObject;
import com.cpf.producer.dao.TransactionLogDao;
import com.cpf.producer.service.OrderService;
import com.cpf.producer.dao.TransactionLogDao;
import com.cpf.producer.entity.OrderBase;
import com.cpf.producer.service.OrderService;
import lombok.extern.slf4j.Slf4j;
import org.apache.rocketmq.client.producer.LocalTransactionState;
import org.apache.rocketmq.client.producer.TransactionListener;
```

```java
import org.apache.rocketmq.common.message.Message;
import org.apache.rocketmq.common.message.MessageExt;
import org.springframework.beans.factory.annotation.Autowired;
import org.springframework.stereotype.Component;

import java.util.concurrent.TimeUnit;

/**
 * 事务监听器
 * @author 晁鹏飞
 */
@Component
@Slf4j
public class OrderTransactionListener implements TransactionListener {

    @Autowired
    OrderService orderService;

    @Autowired
    TransactionLogDao transactionLogDao;

    /**
     *
     * @param message
     * @param o
     * @return
     */
    @Override
    public LocalTransactionState executeLocalTransaction(Message message, Object o) {
        log.info("开始执行本地事务...");
        LocalTransactionState state;
        try{

            String body = new String(message.getBody());
            OrderBase order = JSONObject.parseObject(body, OrderBase.class);
            orderService.createOrder(order,message.getTransactionId());
```

```java
        // 正常情况下，返回 COMMIT_MESSAGE 后，消息能被消费方消费
        state = LocalTransactionState.COMMIT_MESSAGE;
        // 制造一个除 0 异常，在 catch 中，返回 ROLLBACK_MESSAGE, 消费方无法消费
        // System.out.println(1/0);

        // state = LocalTransactionState.UNKNOW;
        // 执行本地事务时程序休眠，返回的是 UNKNOW 状态，消息也无法消费，并且过一段时间还会继续回查
        // TimeUnit.MINUTES.sleep(1);
        log.info("本地事务已提交。{}",message.getTransactionId());
        // state = LocalTransactionState.COMMIT_MESSAGE;


    }catch (Exception e){
        log.info("执行本地事务失败。{}",e);
        state = LocalTransactionState.ROLLBACK_MESSAGE;
    }
    return state;
}


/**
 *
 * @param messageExt
 * @return
 */
@Override
public LocalTransactionState checkLocalTransaction(MessageExt messageExt) {

    // 如果回查发生多次失败，则发邮件提醒，进行人工回查
    log.info("开始回查本地事务状态。{}",messageExt.getTransactionId());
    LocalTransactionState state;
    String transactionId = messageExt.getTransactionId();
    if (transactionLogDao.selectCount(transactionId)>0){
        state = LocalTransactionState.COMMIT_MESSAGE;
    }else {
        state = LocalTransactionState.UNKNOW;
```

```
        }
        System.out.println();
        log.info("结束本地事务状态查询:{}",state);
        return state;
    }
}
```

此处实现 TransactionListener 接口,并实现 executeLocalTransaction(执行本地事务的,一般就是操作 DB 相关内容)和 checkLocalTransaction 方法(用来提供给 broker 进行回查本地事务消息的,把本地事务执行的结果存储到 Redis 或者 DB 中都可以,为回查做数据准备,此时将数据写入 transaction_log 表中,回查时直接查询此表中的信息即可)。

(9)创建事务发送方工具。

在包 com.cpf.producer.config 下创建类 TransactionProducer.java,代码如下。

```
package com.cpf.producer.config;

import org.apache.rocketmq.client.exception.MQClientException;
import org.apache.rocketmq.client.producer.TransactionMQProducer;
import org.apache.rocketmq.client.producer.TransactionSendResult;
import org.apache.rocketmq.common.message.Message;
import org.springframework.beans.factory.annotation.Autowired;
import org.springframework.stereotype.Component;

import javax.annotation.PostConstruct;
import java.util.concurrent.ArrayBlockingQueue;
import java.util.concurrent.ThreadPoolExecutor;
import java.util.concurrent.TimeUnit;

/**
 * 消息发送方工具类
 * @author 晁鹏飞
 */
@Component
public class TransactionProducer {

    private String producerGroup = "order_trans_group";

    // 事务消息
```

```java
private TransactionMQProducer producer;

/**
 * 用于执行本地事务和事务状态回查的监听器
 */
@Autowired
OrderTransactionListener orderTransactionListener;

// 执行任务的线程池
ThreadPoolExecutor executor = new ThreadPoolExecutor(5, 10, 60,
        TimeUnit.SECONDS, new ArrayBlockingQueue<>(50));

@PostConstruct
public void init(){
    producer = new TransactionMQProducer(producerGroup);
    producer.setNamesrvAddr("127.0.0.1:9876");
    producer.setSendMsgTimeout(Integer.MAX_VALUE);
    producer.setExecutorService(executor);
    producer.setTransactionListener(orderTransactionListener);
    this.start();
}

private void start(){
    try {
        this.producer.start();
    } catch (MQClientException e) {
        e.printStackTrace();
    }
}

/**
 * 事务消息发送
 * @param data
 * @param topic
 * @return
 * @throws MQClientException
```

```
    */
    public TransactionSendResult send(String data, String topic) throws
MQClientException {
        Message message = new Message(topic,data.getBytes());
        return this.producer.sendMessageInTransaction(message, null);
    }
}
```

（10）创建 Controller。

在包 com.cpf.producer.controller 下创建类 OrderController.java，代码如下。

```
package com.cpf.producer.controller;

import com.cpf.producer.entity.OrderBase;
import com.cpf.producer.service.OrderService;
import lombok.extern.slf4j.Slf4j;
import org.apache.rocketmq.client.exception.MQClientException;
import org.springframework.beans.factory.annotation.Autowired;
import org.springframework.web.bind.annotation.PostMapping;
import org.springframework.web.bind.annotation.RequestBody;
import org.springframework.web.bind.annotation.RequestMapping;
import org.springframework.web.bind.annotation.RestController;

/**
 * 事务消息 Controller
 * @author 晁鹏飞
 */
@RestController
@Slf4j
@RequestMapping("/order")
public class OrderController {

    @Autowired
    OrderService orderService;

    @PostMapping("/create")
    public void createOrder(@RequestBody OrderBase order) throws
MQClientException {
```

```
        orderService.createOrder(order);
    }
}
```

（11）下面进行测试。

在 Postman 中请求接口 localhost:8081/order/create。method 为 post，body 参数为{ "orderNo": "业务订单号"}。

执行接口请求时会遇到以下 3 种情况。

☑ 本地业务正常执行。

在 executeLocalTransaction 方法中返回 LocalTransactionState.COMMIT_MESSAGE。

```
2021-04-10 01:25:28.272  INFO 35712 --- [nio-8081-exec-2] c.c.p.config.
OrderTransactionListener    : 开始执行本地事务....
2021-04-10 01:25:28.310  INFO 35712 --- [nio-8081-exec-2] com.alibaba.druid.
pool.DruidDataSource   : {dataSource-1} inited
2021-04-10 01:25:28.943  INFO 35712 --- [nio-8081-exec-2] c.c.p.config.
OrderTransactionListener    : 本地事务已提交。C0A86E198B8018B4AAC22EA77C480000
```

观察日志消息发送成功，本地事务也执行成功。

☑ 在本地业务执行时制造一个除 0 异常。

修改 createOrder 方法，在里面制造一个除 0 异常。

```
@Transactional
@Override
public void createOrder(OrderBase order, String transactionId) throws
InvocationTargetException, IllegalAccessException {

    // 1.创建订单
    orderMapper.insert(order);

    // 2.写入事务日志
    TransactionLog log = new TransactionLog();
    log.setId(transactionId);
    log.setBusiness(order.getOrderNo());
    log.setForeignKey(String.valueOf(order.getId()));
    transactionLogMapper.insert(log);

    // 在操作数据库时制造一个除 0 异常
```

```
    System.out.println(1/0);
}
```

请求接口，查看日志。

```
2021-04-10 16:29:11.516  INFO 23052 --- [nio-8081-exec-7] c.c.p.config.
OrderTransactionListener      : 开始执行本地事务....
2021-04-10 16:29:11.528  INFO 23052 --- [nio-8081-exec-7] c.c.p.config.
OrderTransactionListener      : 执行本地事务失败。{}

java.lang.ArithmeticException: / by zero
```

可以看到数据库没有数据，消息队列查看不到消息。

☑ 本地业务超时。

修改 createOrder 方法，在方法增加一个睡眠 5min 的操作，代码如下。

```
@Override
public void createOrder(OrderBase order, String transactionId) throws 
InvocationTargetException, IllegalAccessException {

    // 1.创建订单
    orderMapper.insert(order);

    // 2.写入事务日志
    TransactionLog log = new TransactionLog();
    log.setId(transactionId);
    log.setBusiness(order.getOrderNo());
    log.setForeignKey(String.valueOf(order.getId()));
    transactionLogMapper.insert(log);

    // 在操作数据库时增加一个睡眠 5min 的操作
    try {
        TimeUnit.MINUTES.sleep(5);
    } catch (InterruptedException e) {
        e.printStackTrace();
    }

}
```

请求接口，查看日志。

```
2021-04-10 16:33:36.128  INFO 23052 --- [io-8081-exec-10] c.c.p.config.
OrderTransactionListener        : 开始执行本地事务....
2021-04-10 16:34:15.027  INFO 23052 --- [pool-4-thread-2] c.c.p.config.
OrderTransactionListener        : 开始回查本地事务状态。
C0A86E195A0C18B4AAC231E6E7780006

2021-04-10 16:34:15.032  INFO 23052 --- [pool-4-thread-2] c.c.p.config.
OrderTransactionListener        : 结束本地事务状态查询:ROLLBACK_MESSAGE
2021-04-10 16:35:15.029  INFO 23052 --- [pool-4-thread-3] c.c.p.config.
OrderTransactionListener        : 开始回查本地事务状态。
C0A86E195A0C18B4AAC231E6E7780006

2021-04-10 16:35:15.031  INFO 23052 --- [pool-4-thread-3] c.c.p.config.
OrderTransactionListener        : 结束本地事务状态查询:ROLLBACK_MESSAGE
```

（12）小结。

从上面的代码分析结果得知，通过 RocketMQ 的事务消息，将本地数据库的操作和 MQ 的消息发送放到了同一个事务中，实现了原子操作。

10.10.2 消费者 consumer

consumer 项目的目录结构如图 10-13 所示。

consumer 项目的具体实现步骤如下。

（1）准备数据库。

创建数据库 mq-consumer，在其中创建表 tbl_points，创建表的具体语句如下。

```sql
CREATE TABLE `tbl_points` (
  `id` bigint(16) NOT NULL AUTO_INCREMENT COMMENT '主键',
  `user_id` bigint(16) NOT NULL COMMENT '用户id',
  `order_no` varchar(16) COLLATE utf8mb4_bin NOT NULL COMMENT '订单编号',
  `points` int(4) NOT NULL COMMENT '积分',
  `remarks` varchar(128) CHARACTER SET utf8mb4 COLLATE utf8mb4_bin NOT NULL COMMENT '备注',
  PRIMARY KEY (`id`)
) ENGINE=InnoDB AUTO_INCREMENT=12 DEFAULT CHARSET=utf8mb4 COLLATE=utf8mb4_bin;
```

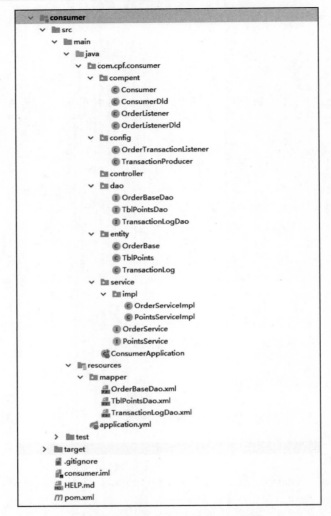

图 10-13　consumer 项目的目录结构

（2）修改项目的 pom 文件。

```
<?xml version="1.0" encoding="UTF-8"?>
<project xmlns="http://maven.apache.org/POM/4.0.0" xmlns:xsi="http://www.w3.org/2001/XMLSchema-instance"
    xsi:schemaLocation="http://maven.apache.org/POM/4.0.0 https://maven.apache.org/xsd/maven-4.0.0.xsd">
    <modelVersion>4.0.0</modelVersion>
```

```xml
<parent>
    <groupId>org.springframework.boot</groupId>
    <artifactId>spring-boot-starter-parent</artifactId>
    <version>2.3.3.RELEASE</version>
    <relativePath/> <!-- lookup parent from repository -->
</parent>
<groupId>com.cpf.consumer</groupId>
<artifactId>consumer</artifactId>
<version>0.0.1-SNAPSHOT</version>
<name>consumer</name>
<description>Demo project for Spring Boot</description>

<properties>
    <java.version>1.8</java.version>
</properties>

<dependencies>
    <!--rocketmq-->
    <dependency>
        <groupId>org.apache.rocketmq</groupId>
        <artifactId>rocketmq-spring-boot-starter</artifactId>
        <version>2.0.2</version>
    </dependency>

    <!-- json-lib -->
    <dependency>
        <groupId>net.sf.json-lib</groupId>
        <artifactId>json-lib</artifactId>
        <version>2.4</version>
        <classifier>jdk15</classifier>
    </dependency>


    <dependency>
        <groupId>org.springframework.boot</groupId>
        <artifactId>spring-boot-starter-web</artifactId>
    </dependency>
```

```xml
<dependency>
    <groupId>org.springframework.boot</groupId>
    <artifactId>spring-boot-starter-test</artifactId>
    <scope>test</scope>
    <exclusions>
        <exclusion>
            <groupId>org.junit.vintage</groupId>
            <artifactId>junit-vintage-engine</artifactId>
        </exclusion>
    </exclusions>
</dependency>

<dependency>
    <groupId>org.springframework</groupId>
    <artifactId>spring-tx</artifactId>
    <version>5.2.8.RELEASE</version>
</dependency>

<dependency>
    <groupId>org.springframework.boot</groupId>
    <artifactId>spring-boot-starter-web</artifactId>
</dependency>

<!-- MyBatis 相关依赖 -->
<dependency>
    <groupId>org.mybatis.spring.boot</groupId>
    <artifactId>mybatis-spring-boot-starter</artifactId>
    <version>2.0.0</version>
</dependency>

<!-- MySQL 驱动 -->
<dependency>
    <groupId>mysql</groupId>
    <artifactId>mysql-connector-java</artifactId>
</dependency>
```

```xml
<!-- 阿里巴巴数据库连接池 -->
<dependency>
    <groupId>com.alibaba</groupId>
    <artifactId>druid</artifactId>
    <version>1.1.12</version>
</dependency>

<dependency>
    <groupId>org.projectlombok</groupId>
    <artifactId>lombok</artifactId>
    <optional>true</optional>
</dependency>


<dependency>
    <groupId>org.springframework.boot</groupId>
    <artifactId>spring-boot-starter</artifactId>
</dependency>

<dependency>
    <groupId>org.springframework.boot</groupId>
    <artifactId>spring-boot-starter-test</artifactId>
    <scope>test</scope>
    <exclusions>
        <exclusion>
            <groupId>org.junit.vintage</groupId>
            <artifactId>junit-vintage-engine</artifactId>
        </exclusion>
    </exclusions>
</dependency>
</dependencies>

<build>
    <plugins>
        <plugin>
            <groupId>org.springframework.boot</groupId>
            <artifactId>spring-boot-maven-plugin</artifactId>
```

```
            </plugin>
        </plugins>
    </build>

</project>
```

（3）修改项目的 application.yml 文件。

```
server:
  port: 8082

spring:
  application:
    name: consumer
  datasource:
    type: com.alibaba.druid.pool.DruidDataSource
    driver-class-name: com.mysql.cj.jdbc.Driver
    url: jdbc:mysql://localhost:3306/mq-consumer?characterEncoding=UTF-8&serverTimezone=Asia/Shanghai
    username: root
    password: root
    dbcp2:
      initial-size: 5
      min-idle: 5
      max-total: 5
      max-wait-millis: 200
      validation-query: SELECT 1
      test-while-idle: true
      test-on-borrow: false
      test-on-return: false

mybatis:
  mapper-locations:
    - classpath:mapper/*.xml
```

（4）编写 mapper 文件。

在 resources 目录下创建目录 mapper。编写访问数据库的 mapper 文件，即表 tbl_points 的操作文件 TblPointsDao.xml，代码如下。

```xml
<?xml version="1.0" encoding="UTF-8"?>
<!DOCTYPE mapper PUBLIC "-//mybatis.org//DTD Mapper 3.0//EN" "http://mybatis.org/dtd/mybatis-3-mapper.dtd">
<mapper namespace="com.cpf.consumer.dao.TblPointsDao">
  <resultMap id="BaseResultMap" type="com.cpf.consumer.entity.TblPoints">
    <id column="id" jdbcType="BIGINT" property="id" />
    <result column="user_id" jdbcType="BIGINT" property="userId" />
    <result column="order_no" jdbcType="VARCHAR" property="orderNo" />
    <result column="points" jdbcType="INTEGER" property="points" />
    <result column="remarks" jdbcType="VARCHAR" property="remarks" />
  </resultMap>
  <sql id="Base_Column_List">
    id, user_id, order_no, points, remarks
  </sql>
  <select id="selectByPrimaryKey" parameterType="java.lang.Long" resultMap="BaseResultMap">
    select
    <include refid="Base_Column_List" />
    from tbl_points
    where id = #{id,jdbcType=BIGINT}
  </select>
  <delete id="deleteByPrimaryKey" parameterType="java.lang.Long">
    delete from tbl_points
    where id = #{id,jdbcType=BIGINT}
  </delete>
  <insert id="insert" keyColumn="id" keyProperty="id" parameterType="com.cpf.consumer.entity.TblPoints" useGeneratedKeys="true">
    insert into tbl_points (user_id, order_no, points,
      remarks)
    values (#{userId,jdbcType=BIGINT}, #{orderNo,jdbcType=VARCHAR}, #{points,jdbcType=INTEGER},
      #{remarks,jdbcType=VARCHAR})
  </insert>
  <insert id="insertSelective" keyColumn="id" keyProperty="id" parameterType="com.cpf.consumer.entity.TblPoints" useGeneratedKeys="true">
    insert into tbl_points
    <trim prefix="(" suffix=")" suffixOverrides=",">
```

```xml
        <if test="userId != null">
          user_id,
        </if>
        <if test="orderNo != null">
          order_no,
        </if>
        <if test="points != null">
          points,
        </if>
        <if test="remarks != null">
          remarks,
        </if>
    </trim>
    <trim prefix="values (" suffix=")" suffixOverrides=",">
        <if test="userId != null">
          #{userId,jdbcType=BIGINT},
        </if>
        <if test="orderNo != null">
          #{orderNo,jdbcType=VARCHAR},
        </if>
        <if test="points != null">
          #{points,jdbcType=INTEGER},
        </if>
        <if test="remarks != null">
          #{remarks,jdbcType=VARCHAR},
        </if>
    </trim>
  </insert>
  <update id="updateByPrimaryKeySelective" parameterType="com.cpf.consumer.entity.TblPoints">
    update tbl_points
    <set>
        <if test="userId != null">
          user_id = #{userId,jdbcType=BIGINT},
        </if>
        <if test="orderNo != null">
          order_no = #{orderNo,jdbcType=VARCHAR},
```

```xml
      </if>
      <if test="points != null">
        points = #{points,jdbcType=INTEGER},
      </if>
      <if test="remarks != null">
        remarks = #{remarks,jdbcType=VARCHAR},
      </if>
    </set>
    where id = #{id,jdbcType=BIGINT}
  </update>
  <update id="updateByPrimaryKey" parameterType="com.cpf.consumer.entity.TblPoints">
    update tbl_points
    set user_id = #{userId,jdbcType=BIGINT},
      order_no = #{orderNo,jdbcType=VARCHAR},
      points = #{points,jdbcType=INTEGER},
      remarks = #{remarks,jdbcType=VARCHAR}
    where id = #{id,jdbcType=BIGINT}
  </update>
</mapper>
```

（5）编写实体类。

在 com.cpf.consumer.entity 包下，创建类 TblPoints.java，代码如下。

```java
package com.cpf.consumer.entity;

import java.io.Serializable;
import lombok.Data;

/**
 * tbl_points
 * @author 晁鹏飞
 */
@Data
public class TblPoints implements Serializable {
    /**
     * 主键
     */
```

```
    private Long id;

    /**
    * 用户id
    */
    private Long userId;

    /**
    * 订单编号
    */
    private String orderNo;

    /**
    * 积分
    */
    private Integer points;

    /**
    * 备注
    */
    private String remarks;

}
```

(6) 编写对应的 DAO 操作类。

在 com.cpf.consumer.dao 包下,创建类 TblPointsDao,代码如下。

```
package com.cpf.consumer.dao;

import com.cpf.consumer.entity.TblPoints;
import com.cpf.consumer.entity.TblPoints;
import org.apache.ibatis.annotations.Mapper;

@Mapper
public interface TblPointsDao {
    int deleteByPrimaryKey(Long id);

    int insert(TblPoints record);
```

```
    int insertSelective(TblPoints record);

    TblPoints selectByPrimaryKey(Long id);

    int updateByPrimaryKeySelective(TblPoints record);

    int updateByPrimaryKey(TblPoints record);
}
```

（7）编写 Service 层。

创建包 com.cpf.consumer.service，定义接口 PointsService，代码如下。

```
package com.cpf.consumer.service;

import com.cpf.consumer.entity.OrderBase;

/**
 * 积分接口
 * @author 晁鹏飞
 */
public interface PointsService {

    public void increasePoints(OrderBase order);
}
```

在包 com.cpf.consumer.service 下创建子包 impl，定义接口 PointsService 的实现类 PointsServiceImpl.java，代码如下。

```
package com.cpf.consumer.service.impl;

import com.cpf.consumer.dao.TblPointsDao;
import com.cpf.consumer.entity.OrderBase;
import com.cpf.consumer.entity.TblPoints;
import com.cpf.consumer.service.PointsService;
import lombok.extern.slf4j.Slf4j;
import org.springframework.beans.factory.annotation.Autowired;
import org.springframework.stereotype.Service;
```

```java
/**
 * 积分接口实现类
 * @author 晁鹏飞
 */
@Service
@Slf4j
public class PointsServiceImpl implements PointsService {

    @Autowired
    TblPointsDao pointsMapper;

    @Override
    public void increasePoints(OrderBase order) {

        TblPoints points = new TblPoints();
        points.setUserId(1L);
        points.setOrderNo("wo");
        points.setPoints(10);
        points.setRemarks("商品消费共 10 元，获得积分" + points.getPoints());
        pointsMapper.insert(points);
        log.info("已为订单号码{}增加积分。", points.getOrderNo());

    }
}
```

（8）定义消息监听器。

创建包 com.cpf.consumer.compent，在其中定义类 OrderListener.java，代码如下。

```java
package com.cpf.consumer.compent;

import com.alibaba.fastjson.JSONObject;
import com.cpf.consumer.service.PointsService;
import com.cpf.consumer.entity.OrderBase;
import com.cpf.consumer.service.PointsService;
import lombok.extern.slf4j.Slf4j;
import org.apache.rocketmq.client.consumer.listener.ConsumeConcurrentlyContext;
```

```java
import org.apache.rocketmq.client.consumer.listener.ConsumeConcurrentlyStatus;
import org.apache.rocketmq.client.consumer.listener.MessageListenerConcurrently;
import org.apache.rocketmq.common.message.MessageExt;
import org.springframework.beans.factory.annotation.Autowired;
import org.springframework.stereotype.Component;

import java.util.List;

/**
 * 消息监听器
 * @author 晁鹏飞
 */
@Component
@Slf4j
public class OrderListener implements MessageListenerConcurrently {

    @Autowired
    PointsService pointsService;

    @Override
    public ConsumeConcurrentlyStatus consumeMessage(List<MessageExt> list, ConsumeConcurrentlyContext context) {
        log.info("消费者线程监听到消息。");
        try{

            //System.out.println(1/0);
            for (MessageExt message:list) {
                log.info("开始处理订单数据,准备增加积分...");
                OrderBase order = JSONObject.parseObject(message.getBody(), OrderBase.class);
                pointsService.increasePoints(order);
            }
            return ConsumeConcurrentlyStatus.CONSUME_SUCCESS;
        }catch (Exception e){
            log.error("处理消费者数据发生异常。{}",e);
```

```
            return ConsumeConcurrentlyStatus.RECONSUME_LATER;
        }
    }
}
```

在此类中发生异常时，会进行重复消费。

（9）创建消息队列消费端。

在包 com.cpf.consumer.compent 下，创建类 Consumer.java，代码如下。

```java
package com.cpf.consumer.compent;

import org.apache.rocketmq.client.consumer.DefaultMQPushConsumer;
import org.apache.rocketmq.client.exception.MQClientException;
import org.springframework.beans.factory.annotation.Autowired;
import org.springframework.stereotype.Component;

import javax.annotation.PostConstruct;

/**
 * 消费端定义
 * @author 晁鹏飞
 */
@Component
public class Consumer {

    String consumerGroup = "consumer-group";
    DefaultMQPushConsumer consumer;

    @Autowired
    OrderListener orderListener;

    @PostConstruct
    public void init() throws MQClientException {
        consumer = new DefaultMQPushConsumer(consumerGroup);
        consumer.setNamesrvAddr("127.0.0.1:9876");
        consumer.subscribe("order-create","*");
        consumer.registerMessageListener(orderListener);
```

```
        // 两次失败，就进死信队列
        consumer.setMaxReconsumeTimes(2);
        consumer.start();
    }
}
```

（10）启动项目 consumer。

发起 producer 的请求，观察 consumer 控制台日志。

```
2021-04-10 18:04:05.342  INFO 23296 --- [MessageThread_1] com.cpf.
consumer.compent.OrderListener   : 消费者线程监听到消息。
2021-04-10 18:04:05.342  INFO 23296 --- [MessageThread_1] com.cpf.
consumer.compent.OrderListener   : 开始处理订单数据，准备增加积分....
2021-04-10 18:04:05.382  INFO 23296 --- [MessageThread_1] com.alibaba.
druid.pool.DruidDataSource       : {dataSource-1} inited
2021-04-10 18:04:05.976  INFO 23296 --- [MessageThread_1] c.c.c.service.
impl.PointsServiceImpl           : 已为订单号码业务订单号增加积分。
```

观察表 tbl_points 数据发现，多了一条积分记录。

（11）死信队列消费。

当消息正常被消费时，一切都正常，而当消息消费异常时，会进行消费重试。如果消费消息对程序一直报错，无法正常消费呢？此时可以用到死信队列。

在 Consumer.java 中设置如果两次消费失败，则将消息放进死信队列。

```
// 两次失败，就进死信队列
consumer.setMaxReconsumeTimes(2);
consumer.start();
```

死信队列的消息，其实也是正常的消息，它可以对死信队列进行监听。

配置死信队列消费端，代码如下。

```
package com.cpf.consumer.compent;

import org.apache.rocketmq.client.consumer.DefaultMQPushConsumer;
import org.apache.rocketmq.client.exception.MQClientException;
import org.springframework.beans.factory.annotation.Autowired;
import org.springframework.stereotype.Component;

import javax.annotation.PostConstruct;
```

```java
/**
 * 死信队列消费端
 * @author 晁鹏飞
 */
@Component
public class ConsumerDld {

    String consumerGroup = "consumer-group1";
    DefaultMQPushConsumer consumer;

    @Autowired
    OrderListenerDld orderListener;

    @PostConstruct
    public void init() throws MQClientException {
        consumer = new DefaultMQPushConsumer(consumerGroup);
        consumer.setNamesrvAddr("127.0.0.1:9876");
        consumer.subscribe("%DLQ%consumer-group","*");
        consumer.registerMessageListener(orderListener);

        consumer.setMaxReconsumeTimes(2);
        consumer.start();
    }
}
```

编写死信队列的监听器 OrderListenerDld.java，代码如下。

```
package com.cpf.consumer.compent;

import com.cpf.consumer.service.PointsService;
import com.cpf.consumer.entity.OrderBase;
import com.cpf.consumer.service.PointsService;
import lombok.extern.slf4j.Slf4j;
import org.apache.rocketmq.client.consumer.listener.
ConsumeConcurrentlyContext;
import org.apache.rocketmq.client.consumer.listener.
ConsumeConcurrentlyStatus;
```

```java
import org.apache.rocketmq.client.consumer.listener.MessageListenerConcurrently;
import org.apache.rocketmq.common.message.MessageExt;
import org.springframework.beans.factory.annotation.Autowired;
import org.springframework.stereotype.Component;

import java.util.List;

/**
 * 死信队列监听器
 * @author 晁鹏飞
 */
@Component
@Slf4j
public class OrderListenerDld implements MessageListenerConcurrently {

    @Autowired
    PointsService pointsService;

    @Override
    public ConsumeConcurrentlyStatus consumeMessage(List<MessageExt> list, ConsumeConcurrentlyContext context) {
        log.info("死信队列：消费者线程监听到消息。");
        try{
            // System.out.println(1/0);
            for (MessageExt message:list) {
                log.info("死信队列：开始处理订单数据，准备增加积分...");

                System.out.println("发邮件");

            }
            return ConsumeConcurrentlyStatus.CONSUME_SUCCESS;
        }catch (Exception e){
            log.error("死信队列：处理消费者数据发生异常。{}",e);
            return ConsumeConcurrentlyStatus.RECONSUME_LATER;
        }
    }
}
```

此时程序对死信队列进行订阅，收到死信队列中的消息后发出通知，可以发短信、邮件等。

10.11　Seata 分布式事务解决方案

　　Seata 是一款开源的分布式事务解决方案，致力于提供高性能和简单易用的分布式事务服务。Seata 为用户提供了 AT、TCC、SAGA 和 XA 事务模式，为用户打造一站式的分布式解决方案。Seata 的官网是 http://seata.io/。

　　在两阶段提交协议中，每个事务参与方在执行全局事务过程中，会占用数据库连接资源。在 Seata 中却解决了这个问题，它通过对事务参与者在应用层面的协调，来完成分布式事务。

　　Seata 和传统的两阶段提交还是有一些区别的。下面我们一起来看 Seata 用到的一些概念组件，如图 10-14 所示。

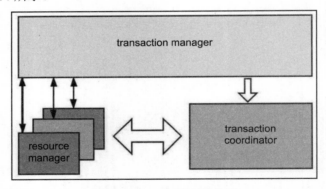

图 10-14　Seata 概念组件图

- ☑　TM（transaction manager）：事务管理器，TM 需要嵌入应用程序，它负责开启一个全局事务，并定义全局事务的范围，它的目的是最终向 TC 发起全局提交或回滚指令。
- ☑　TC（transaction coordinator）：事务协调器，它是独立的组件，需要独立部署运行，Seata 提供了这个独立运行的程序，后面会有实战，它负责维护全局事务的运行状态，接收 TM 指令发起全局事务的提交与回滚，负责与 RM 通信，协调各个分支事务的提交或回滚。
- ☑　RM（resource manager）：与 TC 通信，控制分支事务，负责分支注册、报告分支事务状态，并接收事务协调器 TC 的指令，命令分支事务完成本地事务的提交或回滚。

下面以 Seata 的 AT 模式为例，分析 Seata 作为分布式解决方案的具体流程。假设用户请求订单服务完成订单支付，然后调用积分服务向用户增加积分，Seata 的 AT 模式流程图如图 10-15 所示。

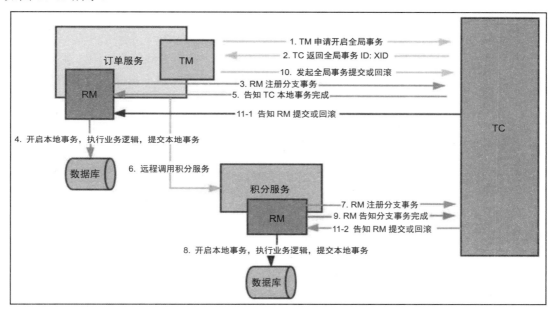

图 10-15　Seata AT 模式流程图

Seata AT 模式的具体流程如下。

（1）订单服务收到请求，订单服务中的 TM 向 TC 申请开启一个全局事务。

（2）TC 收到请求，创建一个全局事务，并将全局事务 ID（称为 XID）返回给订单服务的 TM。

（3）订单服务的 RM 向 TC 注册分支事务。

（4）订单服务执行本地分支事务的业务逻辑并提交，释放锁定的数据库资源。

（5）订单服务向 TC 上报本地分支事务的提交结果。

（6）订单服务调用远程的积分服务，此时将 XID 通过参数传给积分服务。

（7）积分服务向 TC 注册分支事务。

（8）积分服务执行本地分支事务的业务逻辑并提交，释放锁定的数据库资源，并返回订单服务。

（9）积分服务向 TC 上报本地分支事务的提交结果。

（10）订单服务的 TM 向 TC 发起全局事务的提交或回滚。

（11）TC 向 XID 管辖下的全部分支事务发出提交或回滚的指令。

从上面的流程看到，Seata 的 AT 模式和两阶段提交协议是有区别的，如下所示。

- ☑ 设计方面：传统两阶段提交协议方案中的 RM 实际上是在数据库层，RM 本质就是数据库，通过 XA 协议实现，而 Seata 的 RM 是以引入 jar 包的方式实现的，它作为中间件部署在应用程序这一侧。
- ☑ 传统的两阶段提交协议，对数据库资源的占用，需要等到第二阶段才能释放。而 Seata 在第一阶段就将本地事务提交，并释放了资源，从而提高了资源效率。

10.12　Seata AT 模式实战

本节用一个实际的业务例子来进行实战。业务需求如下：有两个服务，一个是订单服务，另一个是库存服务，订单服务接收用户请求，然后调用库存服务进行库存的扣减，扣减完成后，订单服务创建一个订单。分布式事务实战业务流程如图 10-16 所示。

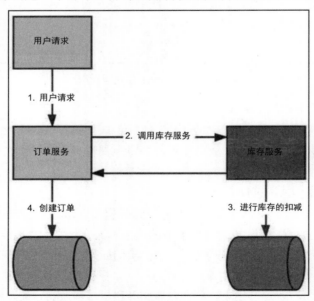

图 10-16　分布式事务实战业务流程图

10.12.1 启动注册中心

启动一个 Eureka 服务端项目 eureka-server，注册中心启动完成如图 10-17 所示。

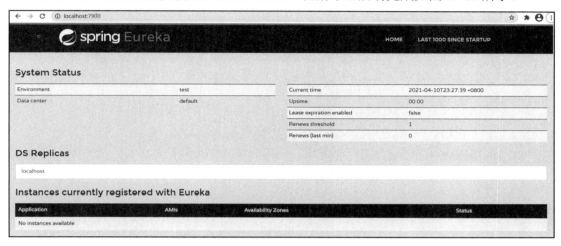

图 10-17　注册中心启动完成

10.12.2 下载安装 Seata

接下来搭建一个 TC，其实 Seata 已经开发好了这个组件，我们直接使用即可。

打开下载地址 https://github.com/seata/seata/releases/download/v1.4.0/seata-server-1.4.0.zip 进行下载，得到一个文件 seata-server-1.4.0.zip，解压后得到目录 seata。

具体操作步骤如下。

（1）配置 registry.conf。

打开 seata/conf 目录下的 registry.conf 文件，修改注册中心为 eureka，并配置 eureka 地址。

```
registry {
  # file 、nacos 、eureka、redis、zk、consul、etcd3、sofa
  # 修改此处为 eureka
  type = "eureka"
  loadBalance = "RandomLoadBalance"
  loadBalanceVirtualNodes = 10
```

```
nacos {
  application = "seata-server"
  serverAddr = "127.0.0.1:8848"
  group = "SEATA_GROUP"
  namespace = ""
  cluster = "default"
  username = ""
  password = ""
}
eureka {
  # 配置eureka地址
  serviceUrl = "http://localhost:7900/eureka"
  application = "default"
  weight = "1"
}
redis {
  serverAddr = "localhost:6379"
  db = 0
  password = ""
  cluster = "default"
  timeout = 0
}
zk {
  cluster = "default"
  serverAddr = "127.0.0.1:2181"
  sessionTimeout = 6000
  connectTimeout = 2000
  username = ""
  password = ""
}
consul {
  cluster = "default"
  serverAddr = "127.0.0.1:8500"
}
etcd3 {
  cluster = "default"
```

```
    serverAddr = "http://localhost:2379"
  }
  sofa {
    serverAddr = "127.0.0.1:9603"
    application = "default"
    region = "DEFAULT_ZONE"
    datacenter = "DefaultDataCenter"
    cluster = "default"
    group = "SEATA_GROUP"
    addressWaitTime = "3000"
  }
  file {
    name = "file.conf"
  }
}

config {
  # file、nacos 、apollo、zk、consul、etcd3
  type = "file"

  nacos {
    serverAddr = "127.0.0.1:8848"
    namespace = ""
    group = "SEATA_GROUP"
    username = ""
    password = ""
  }
  consul {
    serverAddr = "127.0.0.1:8500"
  }
  apollo {
    appId = "seata-server"
    apolloMeta = "http://192.168.1.204:8801"
    namespace = "application"
    apolloAccesskeySecret = ""
  }
```

```
zk {
  serverAddr = "127.0.0.1:2181"
  sessionTimeout = 6000
  connectTimeout = 2000
  username = ""
  password = ""
}
etcd3 {
  serverAddr = "http://localhost:2379"
}
file {
  name = "file.conf"
}
}
```

（2）配置 file.conf。

```
service {
  #transaction service group mapping
  # 修改点1
  #vgroup_mapping.fbs_tx_group = "default"
  vgroup_mapping.my_tx_group = "seata-server"
  #only support when registry.type=file, please don't set multiple addresses

  # 修改点2
  #default.grouplist = "127.0.0.1:8091"
  seata-server.grouplist = "127.0.0.1:8091"
  #disable seata
  disableGlobalTransaction = false
}

## transaction log store, only used in seata-server
store {
  ## store mode: file、db
  # 修改
  mode = "db"
```

```
## file store property
file {
  ## store location dir
  dir = "sessionStore"
}

## database store property
# 修改
db {
  ## the implement of javax.sql.DataSource, such as DruidDataSource
(druid)/BasicDataSource(dbcp) etc.

  datasource = "druid"
  ## mysql/oracle/h2/oceanbase etc.
  db-type = "mysql"
  driver-class-name = "com.mysql.cj.jdbc.Driver"
  url = "jdbc:mysql://127.0.0.1:3306/seata-server?useUnicode=true&useSSL=false&characterEncoding=utf8&serverTimezone=Asia/Shanghai"
  user = "root"
  password = "root"
}
}
```

（3）创建数据库 seata-server。

此时创建的数据库和步骤（2）中的 file.conf 文件中配置的数据库名对应。

在此库中创建 3 张表，即分支事务表 branch_table、全局事务表 global_table、全局锁表 lock_table。

分支事务表 branch_table 的建表 SQL 语句如下。

```
CREATE TABLE `branch_table` (
  `branch_id` bigint(20) NOT NULL,
  `xid` varchar(128) NOT NULL,
  `transaction_id` bigint(20) DEFAULT NULL,
  `resource_group_id` varchar(32) DEFAULT NULL,
  `resource_id` varchar(256) DEFAULT NULL,
  `branch_type` varchar(8) DEFAULT NULL,
```

```
  `status` tinyint(4) DEFAULT NULL,
  `client_id` varchar(64) DEFAULT NULL,
  `application_data` varchar(2000) DEFAULT NULL,
  `gmt_create` datetime(6) DEFAULT NULL,
  `gmt_modified` datetime(6) DEFAULT NULL,
  PRIMARY KEY (`branch_id`),
  KEY `idx_xid` (`xid`)
) ENGINE=InnoDB DEFAULT CHARSET=utf8;
```

全局事务表 global_table 的建表 SQL 语句如下。

```
CREATE TABLE `global_table` (
  `xid` varchar(128) NOT NULL,
  `transaction_id` bigint(20) DEFAULT NULL,
  `status` tinyint(4) NOT NULL,
  `application_id` varchar(32) DEFAULT NULL,
  `transaction_service_group` varchar(32) DEFAULT NULL,
  `transaction_name` varchar(128) DEFAULT NULL,
  `timeout` int(11) DEFAULT NULL,
  `begin_time` bigint(20) DEFAULT NULL,
  `application_data` varchar(2000) DEFAULT NULL,
  `gmt_create` datetime DEFAULT NULL,
  `gmt_modified` datetime DEFAULT NULL,
  PRIMARY KEY (`xid`),
  KEY `idx_gmt_modified_status` (`gmt_modified`,`status`),
  KEY `idx_transaction_id` (`transaction_id`)
) ENGINE=InnoDB DEFAULT CHARSET=utf8;
```

全局锁表 lock_table 的建表 SQL 语句如下。

```
CREATE TABLE `lock_table` (
  `row_key` varchar(128) NOT NULL,
  `xid` varchar(96) DEFAULT NULL,
  `transaction_id` bigint(20) DEFAULT NULL,
  `branch_id` bigint(20) NOT NULL,
  `resource_id` varchar(256) DEFAULT NULL,
  `table_name` varchar(32) DEFAULT NULL,
  `pk` varchar(36) DEFAULT NULL,
```

```
`gmt_create` datetime DEFAULT NULL,
`gmt_modified` datetime DEFAULT NULL,
PRIMARY KEY (`row_key`),
KEY `idx_branch_id` (`branch_id`)
) ENGINE=InnoDB DEFAULT CHARSET=utf8;
```

（4）添加 MySQL 驱动 jar 包。

在 seata/lib/jdbc 目录中有两个 jdbc 驱动 jar 包，即 mysql-connector-java-5.1.35.jar 和 mysql-connector-java-8.0.19.jar。复制 mysql-connector-java-8.0.19.jar 到 jdbc 的上层目录（也就是 lib 目录），否则启动 Seata 时会报没有 MySQL 驱动的错误。

在早期的 Seata 版本中，经常由于用户使用的驱动和 Seata 内置的驱动不一致而发生错误，这个版本中的 Seata 将两个驱动都放到了 jdbc 目录下，让用户自己选择。

（5）启动 Seata。

以 Windows 系统下的操作为例，进入 seata/bin 目录下，双击 seata-server.bat 文件。查看注册中心可发现，seata-server 已经注册进来了，如图 10-18 所示。

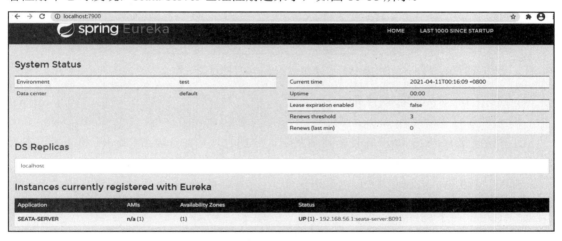

图 10-18　查看注册中心

10.12.3　搭建订单服务

订单服务的功能包括调用库存服务，以及在订单库中创建订单。订单服务 order 项目的目录结构如图 10-19 所示。

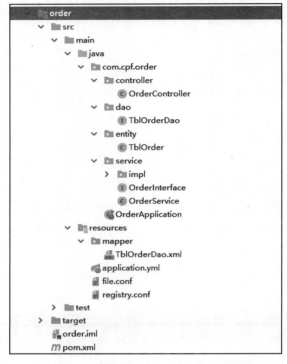

图 10-19　订单服务 order 项目的目录结构

订单服务 order 项目的具体实现步骤如下。

（1）创建订单数据库 seata-order。

创建数据库 seata-order，并且新建两张表，分别为订单业务表 tbl_order 和回滚日志表 undo_log。

订单业务表 tbl_order 的 SQL 建表语句如下。

```sql
CREATE TABLE `tbl_order` (
  `order_id` int(16) NOT NULL AUTO_INCREMENT COMMENT '订单Id',
  `goods_id` int(16) DEFAULT NULL COMMENT '商品ID',
  `buyer` varchar(32) CHARACTER SET utf8 COLLATE utf8_general_ci DEFAULT NULL COMMENT '买家',
  `update_time` timestamp NULL DEFAULT CURRENT_TIMESTAMP ON UPDATE CURRENT_TIMESTAMP COMMENT '更新时间',
  PRIMARY KEY (`order_id`)
) ENGINE=InnoDB AUTO_INCREMENT=56 DEFAULT CHARSET=utf8;
```

回滚日志表 undo_log 用于存储回滚日志，其 SQL 建表语句如下。

```sql
CREATE TABLE `undo_log` (
  `id` bigint(20) NOT NULL AUTO_INCREMENT,
  `branch_id` bigint(20) NOT NULL,
  `xid` varchar(100) NOT NULL,
  `context` varchar(128) NOT NULL,
  `rollback_info` longblob NOT NULL,
  `log_status` int(11) NOT NULL,
  `log_created` datetime NOT NULL,
  `log_modified` datetime NOT NULL,
  `ext` varchar(100) DEFAULT NULL,
  PRIMARY KEY (`id`),
  UNIQUE KEY `ux_undo_log` (`xid`,`branch_id`)
) ENGINE=InnoDB AUTO_INCREMENT=56 DEFAULT CHARSET=utf8;
```

（2）修改 pom 文件，引入相关依赖。

```xml
<?xml version="1.0" encoding="UTF-8"?>
<project xmlns="http://maven.apache.org/POM/4.0.0" xmlns:xsi="http://www.w3.org/2001/XMLSchema-instance"
    xsi:schemaLocation="http://maven.apache.org/POM/4.0.0 https://maven.apache.org/xsd/maven-4.0.0.xsd">

    <modelVersion>4.0.0</modelVersion>

    <parent>
        <groupId>org.springframework.boot</groupId>
        <artifactId>spring-boot-starter-parent</artifactId>
        <version>2.4.2</version>
        <relativePath/> <!-- lookup parent from repository -->
    </parent>

    <groupId>com.cpf</groupId>
    <artifactId>order</artifactId>
    <version>0.0.1-SNAPSHOT</version>
    <name>order</name>
    <description>Demo project for Spring Boot</description>
```

```xml
<properties>
    <java.version>1.8</java.version>
    <spring-cloud.version>2020.0.1</spring-cloud.version>
</properties>

<dependencies>
    <dependency>
        <groupId>org.projectlombok</groupId>
        <artifactId>lombok</artifactId>
        <optional>true</optional>
    </dependency>

    <dependency>
        <groupId>org.springframework.boot</groupId>
        <artifactId>spring-boot-starter-web</artifactId>
    </dependency>
    <dependency>
        <groupId>org.springframework.cloud</groupId>
        <artifactId>spring-cloud-starter-netflix-eureka-client</artifactId>
    </dependency>

    <dependency>
        <groupId>org.springframework.boot</groupId>
        <artifactId>spring-boot-starter-test</artifactId>
        <scope>test</scope>
    </dependency>

    <!-- MyBatis 相关依赖 -->
    <dependency>
        <groupId>org.mybatis.spring.boot</groupId>
        <artifactId>mybatis-spring-boot-starter</artifactId>
        <version>2.0.0</version>
    </dependency>

    <!-- MySQL 驱动 -->
```

```xml
<dependency>
    <groupId>mysql</groupId>
    <artifactId>mysql-connector-java</artifactId>
</dependency>

<!-- 阿里巴巴数据库连接池 -->
<dependency>
    <groupId>com.alibaba</groupId>
    <artifactId>druid</artifactId>
    <version>1.1.12</version>
</dependency>


<!--Seata相关依赖-->
<dependency>
    <groupId>com.alibaba.cloud</groupId>
    <artifactId>spring-cloud-alibaba-seata</artifactId>
    <version>2.2.0.RELEASE</version>
</dependency>

</dependencies>

<dependencyManagement>
    <dependencies>
        <dependency>
            <groupId>org.springframework.cloud</groupId>
            <artifactId>spring-cloud-dependencies</artifactId>
            <version>${spring-cloud.version}</version>
            <type>pom</type>
            <scope>import</scope>
        </dependency>
    </dependencies>
</dependencyManagement>

<build>
    <plugins>
        <plugin>
```

```xml
            <groupId>org.springframework.boot</groupId>
            <artifactId>spring-boot-maven-plugin</artifactId>
        </plugin>
    </plugins>
</build>

</project>
```

其中，重点是引入 Seata 相关依赖。

```xml
<!--Seata 相关依赖-->
<dependency>
    <groupId>com.alibaba.cloud</groupId>
    <artifactId>spring-cloud-alibaba-seata</artifactId>
    <version>2.2.0.RELEASE</version>
</dependency>
```

（3）修改 application.yml。

```yaml
server:
  port: 2001
spring:
  application:
    name: order
  datasource:
    type: com.alibaba.druid.pool.DruidDataSource
    driver-class-name: com.mysql.cj.jdbc.Driver
    url: jdbc:mysql://localhost:3306/seata-order?characterEncoding=UTF-8&serverTimezone=Asia/Shanghai
    username: root
    password: root
    dbcp2:
      initial-size: 5
      min-idle: 5
      max-total: 5
      max-wait-millis: 200
      validation-query: SELECT 1
      test-while-idle: true
      test-on-borrow: false
```

```yaml
      test-on-return: false

mybatis:
  mapper-locations:
    - classpath:mapper/*.xml

eureka:
  client:
    service-url:
      defaultZone: http://localhost:7900/eureka/
```

在 application.yml 文件中配置了注册中心、数据库。

（4）编写 Seata 的配置文件。

在 resources 目录下创建两个文件，即 file.conf 和 registry.conf。

file.conf 文件的内容如下。

```
service {
#   #transaction service group mapping
#   vgroup_mapping.fbs_tx_group= "default"
#   #only support when registry.type=file, please don't set multiple addresses
#   default.grouplist = "127.0.0.1:8091"
#   #disable seata
#   disableGlobalTransaction = false

  #######
  #transaction service group mapping
  vgroup_mapping.my_tx_group="seata-server"
  #only support when registry.type=file, please don't set multiple addresses
  #disable seata
  disableGlobalTransaction = false
}

client {
  rm {
    async.commit.buffer.limit = 10000
    lock {
      retry.internal = 10
```

```
    retry.times = 30
    retry.policy.branch-rollback-on-conflict = true
  }
  report.retry.count = 5
  table.meta.check.enable = false
  report.success.enable = true
}
tm {
  commit.retry.count = 5
  rollback.retry.count = 5
}
undo {
  data.validation = true
  log.serialization = "jackson"
  log.table = "undo_log"
}
log {
  exceptionRate = 100
}
support {
  # auto proxy the DataSource bean
  spring.datasource.autoproxy = false
}
```

vgroup_mapping.my_tx_group="seata-server"配置了 Seata 中间件的服务名（其实是虚拟主机名），用于调用 Seata 服务。

registry.conf 文件的内容如下。

```
registry {
  # file、nacos、eureka、redis、zk、consul、etcd3、sofa
  type = "eureka"

  nacos {
    serverAddr = "localhost"
    namespace = ""
    cluster = "default"
  }
```

```
eureka {
  serviceUrl = "http://localhost:7900/eureka/"
  # 修改点
  application = "seata-server"
  weight = "1"
}
redis {
  serverAddr = "localhost:6379"
  db = "0"
}
zk {
  cluster = "default"
  serverAddr = "127.0.0.1:2181"
  session.timeout = 6000
  connect.timeout = 2000
}
consul {
  cluster = "default"
  serverAddr = "127.0.0.1:8500"
}
etcd3 {
  cluster = "default"
  serverAddr = "http://localhost:2379"
}
sofa {
  serverAddr = "127.0.0.1:9603"
  application = "default"
  region = "DEFAULT_ZONE"
  datacenter = "DefaultDataCenter"
  cluster = "default"
  group = "SEATA_GROUP"
  addressWaitTime = "3000"
}
file {
  name = "file.conf"
}
}
```

```
config {
  # file、nacos、apollo、zk、consul、etcd3
  type = "file"

  nacos {
    serverAddr = "localhost"
    namespace = ""
  }
  consul {
    serverAddr = "127.0.0.1:8500"
  }
  apollo {
    app.id = "seata-server"
    apollo.meta = "http://192.168.1.204:8801"
  }
  zk {
    serverAddr = "127.0.0.1:2181"
    session.timeout = 6000
    connect.timeout = 2000
  }
  etcd3 {
    serverAddr = "http://localhost:2379"
  }
  file {
    name = "file.conf"
  }
}
```

（5）编写 mapper 文件。

在 resources 目录下创建目录 mapper，编写数据库访问的 mapper 文件，即表 tbl_order 的操作文件 TblOrderDao.xml。

```
<?xml version="1.0" encoding="UTF-8"?>
<!DOCTYPE mapper PUBLIC "-//mybatis.org//DTD Mapper 3.0//EN" "http://mybatis.org/dtd/mybatis-3-mapper.dtd">
<mapper namespace="com.cpf.order.dao.TblOrderDao">
  <resultMap id="BaseResultMap" type="com.cpf.order.entity.TblOrder">
```

```xml
    <id column="order_id" jdbcType="INTEGER" property="orderId" />
    <result column="goods_id" jdbcType="VARCHAR" property="goodsId" />
    <result column="buyer" jdbcType="VARCHAR" property="buyer" />
    <result column="update_time" jdbcType="TIMESTAMP" property="updateTime" />
  </resultMap>
  <sql id="Base_Column_List">
    order_id, goods_id, buyer, update_time
  </sql>
  <select id="selectByPrimaryKey" parameterType="java.lang.Integer" resultMap="BaseResultMap">
    select
    <include refid="Base_Column_List" />
    from tbl_order
    where order_id = #{orderId,jdbcType=INTEGER}
  </select>
  <delete id="deleteByPrimaryKey" parameterType="java.lang.Integer">
    delete from tbl_order
    where order_id = #{orderId,jdbcType=INTEGER}
  </delete>
  <insert id="insert" keyColumn="order_id" keyProperty="orderId" parameterType="com.cpf.order.entity.TblOrder" useGeneratedKeys="true">
    insert into tbl_order (buyer, goods_id , update_time)
    values (#{buyer,jdbcType=VARCHAR},#{goodsId}, #{updateTime,jdbcType=TIMESTAMP})
  </insert>
  <insert id="insertSelective" keyColumn="order_id" keyProperty="orderId" parameterType="com.cpf.order.entity.TblOrder" useGeneratedKeys="true">
    insert into tbl_order
    <trim prefix="(" suffix=")" suffixOverrides=",">
      <if test="buyer != null">
        buyer,
      </if>
    <if test="goodsId != null">
        goods_id,
    </if>
      <if test="updateTime != null">
```

```xml
      update_time,
    </if>
  </trim>
  <trim prefix="values (" suffix=")" suffixOverrides=",">
    <if test="buyer != null">
      #{buyer,jdbcType=VARCHAR},
    </if>
      <if test="goodsId != null">
        goodsId,
      </if>
    <if test="updateTime != null">
      #{updateTime,jdbcType=TIMESTAMP},
    </if>
  </trim>
</insert>
<update id="updateByPrimaryKeySelective" parameterType="com.cpf.order.entity.TblOrder">
  update tbl_order
  <set>
    <if test="buyer != null">
      buyer = #{buyer,jdbcType=VARCHAR},
    </if>
      <if test="goodsId != null">
        goods_id = #{goodsId},
      </if>
    <if test="updateTime != null">
      update_time = #{updateTime,jdbcType=TIMESTAMP},
    </if>
  </set>
  where order_id = #{orderId,jdbcType=INTEGER}
</update>
<update id="updateByPrimaryKey" parameterType="com.cpf.order.entity.TblOrder">
  update tbl_order
  set buyer = #{buyer,jdbcType=VARCHAR},
    goods_id = #{goodsId},
    update_time = #{updateTime,jdbcType=TIMESTAMP}
```

```xml
      where order_id = #{orderId,jdbcType=INTEGER}
    </update>
</mapper>
```

（6）编写实体类。

在 com.cpf.order.entity 包中，创建类 TblOrder.java，代码如下。

```java
package com.cpf.order.entity;

import java.io.Serializable;
import java.util.Date;
import lombok.Data;

/**
 * tbl_order 实体类
 * @author 晁鹏飞
 */
@Data
public class TblOrder implements Serializable {
    private Integer orderId;

    private Integer goodsId;

    private String buyer;

    private Date updateTime;

}
```

（7）编写对应的 DAO 操作类。

在 com.cpf.order.dao 包中，创建类 TblOrderDao.java，代码如下。

```java
package com.cpf.order.dao;

import com.cpf.order.entity.TblOrder;
import org.apache.ibatis.annotations.Mapper;

/**
 * 订单数据库操作类
```

```
 * @author 晁鹏飞
 */
@Mapper
public interface TblOrderDao {
    int deleteByPrimaryKey(Integer orderId);

    int insert(TblOrder record);

    int insertSelective(TblOrder record);

    TblOrder selectByPrimaryKey(Integer orderId);

    int updateByPrimaryKeySelective(TblOrder record);

    int updateByPrimaryKey(TblOrder record);
}
```

（8）编写 Service 层。

创建包 com.cpf.order.service，定义服务类 OrderService.java，此时并没有定义服务的接口，而是直接写的实现类，这是为了和后面 TCC 编程模型做对比。

```
package com.cpf.order.service;

import com.cpf.order.entity.TblOrder;
import com.cpf.order.dao.TblOrderDao;
import io.seata.spring.annotation.GlobalTransactional;
import org.springframework.beans.factory.annotation.Autowired;
import org.springframework.stereotype.Service;
import org.springframework.transaction.annotation.Transactional;
import org.springframework.web.client.RestTemplate;


/**
 * 订单服务
 * @author 晁鹏飞
 */
@Service
public class OrderService {
```

```java
    @Autowired
    private RestTemplate restTemplate;

    @Autowired
    TblOrderDao tblOrderDao;

    @GlobalTransactional(rollbackFor = Exception.class)
    @Transactional(rollbackFor = Exception.class)
    public String addOrder(Integer goodsId) {
        restTemplate.getForEntity("http://inventory/reduce?goodsId="+goodsId, null);

        TblOrder tblOrder = new TblOrder();
        tblOrder.setOrderId(1);
        tblOrder.setGoodsId(goodsId);
        tblOrder.setBuyer("晁鹏飞");

        tblOrderDao.insert(tblOrder);

        // 模拟异常
        // System.out.println(1/0);

        return "";
    }
}
```

注意，此时注解@GlobalTransactional(rollbackFor = Exception.class)开启了全局事务，就相当于在订单服务中植入了一个 TM。

为了在调用库存服务时能够通过服务名调用，此时需要引入 RestTemplate。

```java
package com.cpf.order;

import org.springframework.boot.SpringApplication;
import org.springframework.boot.autoconfigure.SpringBootApplication;
import org.springframework.cloud.client.loadbalancer.LoadBalanced;
import org.springframework.context.annotation.Bean;
```

```java
import org.springframework.web.client.RestTemplate;

@SpringBootApplication
public class OrderApplication {

    public static void main(String[] args) {
        SpringApplication.run(OrderApplication.class, args);
    }

    @Bean
    @LoadBalanced
    public RestTemplate restTemplate() {
        return new RestTemplate();
    }
}
```

（9）定义 Controller 层。

创建 Controller 层代码，即 OrderAtController.java，代码如下。

```java
package com.cpf.order.controller;

import com.cpf.order.service.OrderService;
import org.springframework.beans.factory.annotation.Autowired;
import org.springframework.web.bind.annotation.GetMapping;
import org.springframework.web.bind.annotation.RestController;

/**
 * Seata AT 模式 测试控制层
 * @author 晁鹏飞
 */
@RestController
public class OrderAtController {

    @Autowired
    OrderService orderService;

    @GetMapping("/order-add")
    public String addOrder(Integer goodsId) throws InterruptedException {
```

```
        orderService.addOrder(goodsId);
        return "success";
    }

}
```

启动订单服务，如图 10-20 所示。

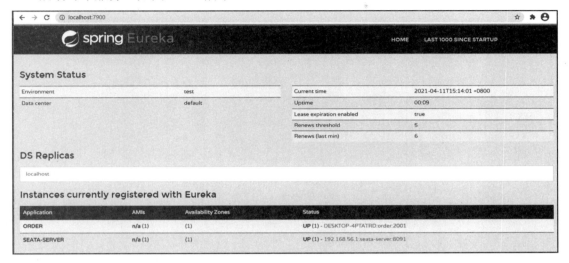

图 10-20　order 订单服务启动

10.12.4　搭建库存服务

库存服务的功能是完成库存的扣减，库存服务 inventory 项目的目录结构如图 10-21 所示。库存服务 inventory 项目的具体实现步骤如下。

（1）创建数据库，并新建两张表。

创建库存数据库 tbl_inventory，并创建库存业务表 tbl_inventory 和回滚日志表 undo_log。库存业务表 tbl_inventory 的 SQL 建表语句如下。

```
CREATE TABLE `tbl_inventory` (
  `good_id` int(16) NOT NULL COMMENT '商品ID',
  `num` int(8) DEFAULT NULL COMMENT '库存数量',
  `update_time` timestamp NULL DEFAULT CURRENT_TIMESTAMP ON UPDATE CURRENT_TIMESTAMP COMMENT '更新时间',
```

```
  PRIMARY KEY (`good_id`)
) ENGINE=InnoDB DEFAULT CHARSET=utf8;
```

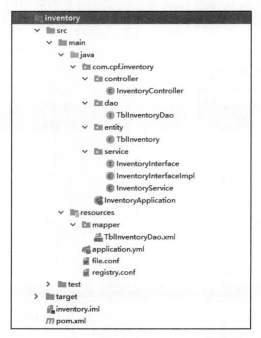

图 10-21 库存服务 inventory 项目的目录结构

回滚日志表 undo_log 用于存储回滚日志，SQL 建表语句如下。

```
CREATE TABLE `undo_log` (
  `id` bigint(20) NOT NULL AUTO_INCREMENT,
  `branch_id` bigint(20) NOT NULL,
  `xid` varchar(100) NOT NULL,
  `context` varchar(128) NOT NULL,
  `rollback_info` longblob NOT NULL,
  `log_status` int(11) NOT NULL,
  `log_created` datetime NOT NULL,
  `log_modified` datetime NOT NULL,
  `ext` varchar(100) DEFAULT NULL,
  PRIMARY KEY (`id`),
  UNIQUE KEY `ux_undo_log` (`xid`,`branch_id`)
) ENGINE=InnoDB AUTO_INCREMENT=56 DEFAULT CHARSET=utf8;
```

（2）修改 pom 文件，引入相关依赖。

```xml
<?xml version="1.0" encoding="UTF-8"?>
<project xmlns="http://maven.apache.org/POM/4.0.0" xmlns:xsi="http://www.w3.org/2001/XMLSchema-instance"
    xsi:schemaLocation="http://maven.apache.org/POM/4.0.0 https://maven.apache.org/xsd/maven-4.0.0.xsd">
    <modelVersion>4.0.0</modelVersion>
    <parent>
        <groupId>org.springframework.boot</groupId>
        <artifactId>spring-boot-starter-parent</artifactId>
        <version>2.4.2</version>
        <relativePath/> <!-- lookup parent from repository -->
    </parent>
    <groupId>com.cpf</groupId>
    <artifactId>inventory</artifactId>
    <version>0.0.1-SNAPSHOT</version>
    <name>inventory</name>
    <description>Demo project for Spring Boot</description>

    <properties>
        <java.version>1.8</java.version>
        <spring-cloud.version>2020.0.1</spring-cloud.version>
    </properties>

    <dependencies>
        <dependency>
            <groupId>org.projectlombok</groupId>
            <artifactId>lombok</artifactId>
            <optional>true</optional>
        </dependency>

        <dependency>
            <groupId>org.springframework.boot</groupId>
            <artifactId>spring-boot-starter-web</artifactId>
        </dependency>
        <dependency>
            <groupId>org.springframework.cloud</groupId>
```

```xml
            <artifactId>spring-cloud-starter-netflix-eureka-client</artifactId>
        </dependency>

        <dependency>
            <groupId>org.springframework.boot</groupId>
            <artifactId>spring-boot-starter-test</artifactId>
            <scope>test</scope>
        </dependency>

        <!-- MyBatis 相关依赖 -->
        <dependency>
            <groupId>org.mybatis.spring.boot</groupId>
            <artifactId>mybatis-spring-boot-starter</artifactId>
            <version>2.0.0</version>
        </dependency>

        <!-- MySQL 驱动 -->
        <dependency>
            <groupId>mysql</groupId>
            <artifactId>mysql-connector-java</artifactId>
        </dependency>

        <!-- 阿里巴巴数据库连接池 -->
        <dependency>
            <groupId>com.alibaba</groupId>
            <artifactId>druid</artifactId>
            <version>1.1.12</version>
        </dependency>

        <!-- Seata 相关依赖-->
        <dependency>
            <groupId>com.alibaba.cloud</groupId>
            <artifactId>spring-cloud-alibaba-seata</artifactId>
            <version>2.2.0.RELEASE</version>
        </dependency>

    </dependencies>
```

```xml
<dependencyManagement>
    <dependencies>
        <dependency>
            <groupId>org.springframework.cloud</groupId>
            <artifactId>spring-cloud-dependencies</artifactId>
            <version>${spring-cloud.version}</version>
            <type>pom</type>
            <scope>import</scope>
        </dependency>
    </dependencies>
</dependencyManagement>

<build>
    <plugins>
        <plugin>
            <groupId>org.springframework.boot</groupId>
            <artifactId>spring-boot-maven-plugin</artifactId>
        </plugin>
    </plugins>
</build>
</project>
```

其中的重点是引入 Seata 的相关依赖。

```xml
<!--Seata 相关依赖-->
<dependency>
    <groupId>com.alibaba.cloud</groupId>
    <artifactId>spring-cloud-alibaba-seata</artifactId>
    <version>2.2.0.RELEASE</version>
</dependency>
```

（3）修改 application.yml。

```yaml
server:
  port: 2002

spring:
```

```yaml
  application:
    name: inventory

  datasource:
    type: com.alibaba.druid.pool.DruidDataSource
    driver-class-name: com.mysql.cj.jdbc.Driver
    url: jdbc:mysql://localhost:3306/seata-inventory?characterEncoding=UTF-8&serverTimezone=Asia/Shanghai
    username: root
    password: root
    dbcp2:
      initial-size: 5
      min-idle: 5
      max-total: 5
      max-wait-millis: 200
      validation-query: SELECT 1
      test-while-idle: true
      test-on-borrow: false
      test-on-return: false

mybatis:
  mapper-locations:
  - classpath:mapper/*.xml

eureka:
  client:
    service-url:
      defaultZone: http://localhost:7900/eureka/
```

在 application.yml 文件中配置了注册中心、数据库。

（4）编写 Seata 的配置文件。

在 resources 目录下创建文件 file.conf 和 registry.conf。

file.conf 文件的内容如下。

```
service {
#  #transaction service group mapping
#  vgroup_mapping.fbs_tx_group= "default"
#  #only support when registry.type=file, please don't set multiple addresses
```

```
#    default.grouplist = "127.0.0.1:8091"
#    #disable seata
#    disableGlobalTransaction = false

    #######
    #transaction service group mapping
    vgroup_mapping.my_tx_group="seata-server"
    #only support when registry.type=file, please don't set multiple addresses
    #disable seata
    disableGlobalTransaction = false
}

client {
  rm {
    async.commit.buffer.limit = 10000
    lock {
      retry.internal = 10
      retry.times = 30
      retry.policy.branch-rollback-on-conflict = true
    }
    report.retry.count = 5
  table.meta.check.enable = false
    report.success.enable = true
  }
  tm {
    commit.retry.count = 5
    rollback.retry.count = 5
  }
  undo {
    data.validation = true
    log.serialization = "jackson"
    log.table = "undo_log"
  }
  log {
    exceptionRate = 100
  }
  support {
    # auto proxy the DataSource bean
```

```
    spring.datasource.autoproxy = false
  }
}
```

vgroup_mapping.my_tx_group="seata-server"配置了 Seata 中间件的服务名（其实是虚拟主机名），用于调用 Seata 服务。

registry.conf 文件的内容如下。

```
registry {
  # file、nacos、eureka、redis、zk、consul、etcd3、sofa
  type = "eureka"

  nacos {
    serverAddr = "localhost"
    namespace = ""
    cluster = "default"
  }
  eureka {
    serviceUrl = "http://localhost:7900/eureka/"
    # 修改点
    application = "seata-server"
    weight = "1"
  }
  redis {
    serverAddr = "localhost:6379"
    db = "0"
  }
  zk {
    cluster = "default"
    serverAddr = "127.0.0.1:2181"
    session.timeout = 6000
    connect.timeout = 2000
  }
  consul {
    cluster = "default"
    serverAddr = "127.0.0.1:8500"
  }
  etcd3 {
```

```
    cluster = "default"
    serverAddr = "http://localhost:2379"
  }
  sofa {
    serverAddr = "127.0.0.1:9603"
    application = "default"
    region = "DEFAULT_ZONE"
    datacenter = "DefaultDataCenter"
    cluster = "default"
    group = "SEATA_GROUP"
    addressWaitTime = "3000"
  }
  file {
    name = "file.conf"
  }
}

config {
  # file、nacos、apollo、zk、consul、etcd3
  type = "file"

  nacos {
    serverAddr = "localhost"
    namespace = ""
  }
  consul {
    serverAddr = "127.0.0.1:8500"
  }
  apollo {
    app.id = "seata-server"
    apollo.meta = "http://192.168.1.204:8801"
  }
  zk {
    serverAddr = "127.0.0.1:2181"
    session.timeout = 6000
    connect.timeout = 2000
  }
```

```
etcd3 {
  serverAddr = "http://localhost:2379"
}
file {
  name = "file.conf"
}
}
```

（5）编写 mapper 文件。

在 resources 目录下创建目录 mapper，编写数据库访问的 mapper 文件，即表 tbl_inventory 的操作文件 TblInventoryDao.xml。

```xml
<?xml version="1.0" encoding="UTF-8"?>
<!DOCTYPE mapper PUBLIC "-//mybatis.org//DTD Mapper 3.0//EN" "http://mybatis.org/dtd/mybatis-3-mapper.dtd">
<mapper namespace="com.cpf.inventory.dao.TblInventoryDao">
  <resultMap id="BaseResultMap" type="com.cpf.inventory.entity.TblInventory">
    <id column="good_id" jdbcType="INTEGER" property="goodId" />
    <result column="num" jdbcType="INTEGER" property="num" />
    <result column="update_time" jdbcType="TIMESTAMP" property="updateTime" />
  </resultMap>
  <sql id="Base_Column_List">
    good_id, num, update_time
  </sql>
  <select id="selectByPrimaryKey" parameterType="java.lang.Integer" resultMap="BaseResultMap">
    select
    <include refid="Base_Column_List" />
    from tbl_inventory
    where good_id = #{goodId,jdbcType=INTEGER}
  </select>
  <delete id="deleteByPrimaryKey" parameterType="java.lang.Integer">
    delete from tbl_inventory
    where good_id = #{goodId,jdbcType=INTEGER}
  </delete>
  <insert id="insert" keyColumn="good_id" keyProperty="goodId"
```

```xml
  parameterType="com.cpf.inventory.entity.TblInventory" useGeneratedKeys="true">
    insert into tbl_inventory (num, update_time)
    values (#{num,jdbcType=INTEGER}, #{updateTime,jdbcType=TIMESTAMP})
  </insert>
  <insert id="insertSelective" keyColumn="good_id" keyProperty="goodId" parameterType="com.cpf.inventory.entity.TblInventory" useGeneratedKeys="true">
    insert into tbl_inventory
    <trim prefix="(" suffix=")" suffixOverrides=",">
      <if test="num != null">
        num,
      </if>
      <if test="updateTime != null">
        update_time,
      </if>
    </trim>
    <trim prefix="values (" suffix=")" suffixOverrides=",">
      <if test="num != null">
        #{num,jdbcType=INTEGER},
      </if>
      <if test="updateTime != null">
        #{updateTime,jdbcType=TIMESTAMP},
      </if>
    </trim>
  </insert>
  <update id="updateByPrimaryKeySelective" parameterType="com.cpf.inventory.entity.TblInventory">
    update tbl_inventory
    <set>
      <if test="num != null">
        num = #{num,jdbcType=INTEGER},
      </if>
      <if test="updateTime != null">
        update_time = #{updateTime,jdbcType=TIMESTAMP},
      </if>
    </set>
    where good_id = #{goodId,jdbcType=INTEGER}
```

```xml
    </update>
    <update id="updateByPrimaryKey" parameterType="com.cpf.inventory.entity.TblInventory">
      update tbl_inventory
      set num = #{num,jdbcType=INTEGER},
        update_time = #{updateTime,jdbcType=TIMESTAMP}
      where good_id = #{goodId,jdbcType=INTEGER}
    </update>
</mapper>
```

（6）编写实体类。

在 com.cpf.inventory.entity 包下，创建类 TblInventory.java，代码如下。

```java
package com.cpf.inventory.entity;

import java.io.Serializable;
import java.util.Date;
import lombok.Data;

/**
 * tbl_inventory 实体类
 * @author  晁鹏飞
 */
@Data
public class TblInventory implements Serializable {
    private Integer goodId;

    private Integer num;

    private Date updateTime;

    private static final long serialVersionUID = 1L;
}
```

（7）编写对应的 DAO 操作类。

在 com.cpf.inventory.dao 包下，创建类 TblInventoryDao.java，代码如下。

```java
package com.cpf.inventory.dao;
```

```
import com.cpf.inventory.entity.TblInventory;
import org.apache.ibatis.annotations.Mapper;

/**
 * 库存数据库操作类
 * @author 晁鹏飞
 */
@Mapper
public interface TblInventoryDao {
    int deleteByPrimaryKey(Integer goodId);

    int insert(TblInventory record);

    int insertSelective(TblInventory record);

    TblInventory selectByPrimaryKey(Integer goodId);

    int updateByPrimaryKeySelective(TblInventory record);

    int updateByPrimaryKey(TblInventory record);
}
```

（8）编写 Service 层。

创建包 com.cpf.inventory.service，定义服务类 InventoryService.java，此时并没有定义服务的接口，而是直接写的实现类，这样做是为了和后面 TCC 编程模型做对比。

```
package com.cpf.inventory.service;

import java.util.Random;

import com.cpf.inventory.dao.TblInventoryDao;
import com.cpf.inventory.entity.TblInventory;
import org.springframework.beans.factory.annotation.Autowired;
import org.springframework.stereotype.Service;
import org.springframework.transaction.annotation.Transactional;


/**
```

```java
 * 库存服务
 * @author 晁鹏飞
 */
@Service
public class InventoryService {


    @Autowired
    TblInventoryDao tblInventoryDao;

    @Transactional
    public String reduce(int goodId) {
        TblInventory tblInventory = tblInventoryDao.selectByPrimaryKey(goodId);
        tblInventory.setNum(tblInventory.getNum()-1);

        tblInventoryDao.updateByPrimaryKey(tblInventory);

        return "";
    }

}
```

注意，此处并不需要注解@GlobalTransactional(rollbackFor = Exception.class)，因为库存服务是被调用方，不是 TM。

（9）定义 Controller 层。

创建 Controller 层代码，即 InventoryAtController.java，代码如下。

```java
package com.cpf.inventory.controller;

import com.cpf.inventory.service.InventoryService;
import com.cpf.inventory.service.InventoryInterface;
import io.seata.spring.annotation.GlobalTransactional;
import org.springframework.beans.factory.annotation.Autowired;
import org.springframework.web.bind.annotation.GetMapping;
import org.springframework.web.bind.annotation.RestController;

/**
```

```
 * Seata AT 模式 测试控制层
 * @author 晁鹏飞
 */
@RestController
public class InventoryAtController {

    @Autowired
    private InventoryService inventoryService;

    @GetMapping("/reduce")
    public String reduce(Integer goodsId){

        inventoryService.reduce(goodsId);
        return "success";
    }
}
```

启动库存服务，如图 10-22 所示。

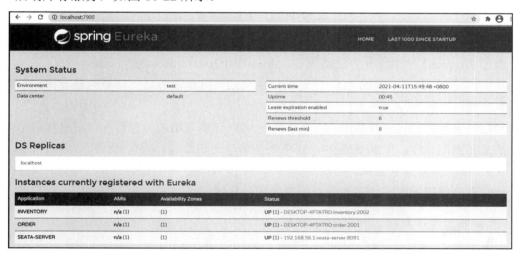

图 10-22　库存服务启动成功

10.12.5　测试

具体的测试步骤如下。

（1）准备初始数据。

在库存库中的库存表中，将 good_id 为 1 的商品库存设置为 100，如图 10-23 所示。

图 10-23　库存表准备数据

订单库中的订单表数据为空，如图 10-24 所示。

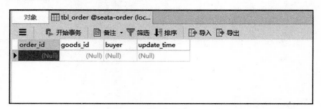

图 10-24　订单表数据准备

（2）请求正常服务测试。

通过 Postman，使用 get 方法请求接口 localhost:2001/order-add?goodsId=1。

其中，goodsId 表示商品 ID，此时的请求会通过订单服务进行订单的创建，订单服务同时也会通过调用库存服务进行库存的扣减。

预期结果是库存表库存减 1，从 100 变成 99，订单表增加一条数据。发送接口请求发现，结果和预期一样：库存变成 99，订单表新增了一条数据。

（3）请求异常服务测试。

此时在订单服务中手动注入一个模拟异常，具体代码如下。

```
package com.cpf.order.service;

import com.cpf.order.entity.TblOrder;
import com.cpf.order.dao.TblOrderDao;
import io.seata.spring.annotation.GlobalTransactional;
import org.springframework.beans.factory.annotation.Autowired;
import org.springframework.stereotype.Service;
import org.springframework.transaction.annotation.Transactional;
```

```java
import org.springframework.web.client.RestTemplate;

/**
 * 订单服务
 * @author 晁鹏飞
 */
@Service
public class OrderService {

    @Autowired
    private RestTemplate restTemplate;

    @Autowired
    TblOrderDao tblOrderDao;

    @GlobalTransactional(rollbackFor = Exception.class)
    @Transactional(rollbackFor = Exception.class)
    public String addOrder(Integer goodsId) {
        restTemplate.getForEntity("http://inventory/reduce?goodsId="+ goodsId, null);

        TblOrder tblOrder = new TblOrder();
        tblOrder.setOrderId(1);
        tblOrder.setGoodsId(goodsId);
        tblOrder.setBuyer("晁鹏飞");

        tblOrderDao.insert(tblOrder);

        // 模拟异常
        System.out.println(1/0);

        return "";
    }
}
```

将库存恢复成 100，将订单表清空后再进行测试。

预期结果是库存表不变，库存还是 100，订单表也不变，订单数据依然为空。发送接口请求后发现结果和预期一样，库存还是 100，订单表无新增数据。

10.13　Seata TCC 模式实战

本节的例子将对 10.12 节的项目进行扩展，编写 TCC 模式的分布式事务解决方案，将通过控制台打印的方式来验证。

10.13.1　订单服务

下面将一步一步完成项目，具体步骤如下。

（1）编写 TCC 服务接口。

在 com.cpf.order.service 包中创建接口 OrderInterface，代码如下。

```java
package com.cpf.order.service;

import io.seata.rm.tcc.api.BusinessActionContext;
import io.seata.rm.tcc.api.LocalTCC;
import io.seata.rm.tcc.api.TwoPhaseBusinessAction;

/**
 * 订单服务 TCC 服务接口
 * @author 晁鹏飞
 */
@LocalTCC
public interface OrderInterface {

    @TwoPhaseBusinessAction(name = "orderTry" , commitMethod = "orderCommit" ,rollbackMethod = "orderRollback")
    public String orderTry(BusinessActionContext businessActionContext);

    public boolean orderCommit(BusinessActionContext businessActionContext);
```

```
    public boolean orderRollback(BusinessActionContext
businessActionContext);
}
```

需要注意以下两点。

① 在接口上添加注解@LocalTCC，表示这个服务是运行在 TCC 编程模型中。该注解需要添加到上面描述的接口上，表示实现该接口的类被 Seata 来管理，Seata 根据事务的状态自动调用定义的方法，如果没问题，则调用 Commit 方法，否则调用 Rollback 方法。

② 在 Try 方法上，添加注解@TwoPhaseBusinessAction(name = "orderTry", commitMethod = "orderCommit",rollbackMethod = "orderRollback")，其中的参数说明如下。

- ☑ name：当前 Try 方法的 Bean 名称，需要全局唯一，一般写方法名即可。
- ☑ commitMethod：Commit 方法的方法名。
- ☑ rollbackMethod：Rollback 方法的方法名。

（2）编写 TCC 实现类。

创建包 com.cpf.order.service.impl，在包中创建类 OrderInterfaceImpl，代码如下。

```
package com.cpf.order.service.impl;

import com.cpf.order.service.OrderInterface;
import io.seata.rm.tcc.api.BusinessActionContext;
import org.springframework.beans.factory.annotation.Autowired;
import org.springframework.stereotype.Component;
import org.springframework.transaction.annotation.Transactional;
import org.springframework.web.client.RestTemplate;

/**
 * TCC 订单实现类
 * @author 晁鹏飞
 */
@Component
public class OrderInterfaceImpl implements OrderInterface {

    /**
     * Try 方法
     * @param businessActionContext
     * @return
```

```java
     */
    @Override
    @Transactional
    public String orderTry(BusinessActionContext businessActionContext) {
        System.out.println("order try");

        inventoryTcc();
        // System.out.println(1/0);
        return null;
    }

    /**
     * Commit 方法
     * @param businessActionContext
     * @return
     */
    @Override
    @Transactional
    public boolean orderCommit(BusinessActionContext businessActionContext) {
        System.out.println("order confirm");
        return true;
    }

    /**
     * Rollback 方法
     * @param businessActionContext
     * @return
     */
    @Override
    @Transactional
    public boolean orderRollback(BusinessActionContext businessActionContext) {
        System.out.println("order cancel");
        return true;
    }
```

```java
    @Autowired
    private RestTemplate restTemplate;

    private void inventoryTcc() {
        restTemplate.getForEntity("http://inventory/inventory-tcc", null);
    }

}
```

在此实现类中,订单服务的 Try 方法调用了库存服务的一个接口。

(3) 编写 TCC 控制类。

创建包 com.cpf.order.controller,在包中创建类 OrderTccController,代码如下。

```java
package com.cpf.order.controller;

import com.cpf.order.service.OrderInterface;
import com.cpf.order.service.OrderService;
import io.seata.spring.annotation.GlobalTransactional;
import org.springframework.beans.factory.annotation.Autowired;
import org.springframework.web.bind.annotation.GetMapping;
import org.springframework.web.bind.annotation.RestController;

/**
 * Seata 测试控制层
 * @author 晁鹏飞
 */
@RestController
public class OrderTccController {

    @Autowired
    private OrderInterface orderInterface;

    @GetMapping("/order-tcc")
    @GlobalTransactional(rollbackFor = Exception.class)
    public String oneTcc() throws InterruptedException {
        orderInterface.orderTry(null);
```

```
        return "success";
    }

}
```

此类调用了步骤（2）中定义的服务。

10.13.2 库存服务

下面完成库存服务项目，具体步骤如下。

（1）编写 TCC 服务接口。

```java
package com.cpf.inventory.service;

import io.seata.rm.tcc.api.BusinessActionContext;
import io.seata.rm.tcc.api.LocalTCC;
import io.seata.rm.tcc.api.TwoPhaseBusinessAction;
/**
 * 库存服务 TCC 服务接口
 * @author 晁鹏飞
 */
@LocalTCC
public interface InventoryInterface {

    @TwoPhaseBusinessAction(name = "inventoryTry" , commitMethod = "inventoryCommit" ,rollbackMethod = "inventoryRollback")
    public String inventoryTry(BusinessActionContext businessActionContext);

    public boolean inventoryCommit(BusinessActionContext businessActionContext);

    public boolean inventoryRollback(BusinessActionContext businessActionContext);
}
```

此接口的定义和订单服务中的类似，都是添加了两个注解 @LocalTCC、@TwoPhaseBusinessAction，并定义好 Try、Commit 和 Rollback 方法。

（2）编写 TCC 实现类。

创建包 com.cpf.inventory.service，在包中创建类 InventoryInterfaceImpl，代码如下。

```java
package com.cpf.inventory.service;

import io.seata.rm.tcc.api.BusinessActionContext;
import org.springframework.beans.factory.annotation.Autowired;
import org.springframework.stereotype.Component;
import org.springframework.transaction.annotation.Transactional;
import org.springframework.web.client.RestTemplate;
/**
 * TCC 库存实现类
 * @author 晁鹏飞
 */
@Component
public class InventoryInterfaceImpl implements InventoryInterface {

    @Override
    @Transactional
    public String inventoryTry(BusinessActionContext businessActionContext) {
        System.out.println("inventory try");
        // System.out.println(1/0);

        return null;
    }

    @Override
    @Transactional
    public boolean inventoryCommit(BusinessActionContext businessActionContext) {
        System.out.println("inventory confirm");
        return true;
    }

    @Override
    @Transactional
    public boolean inventoryRollback(BusinessActionContext businessActionContext) {
```

```
            System.out.println("inventory cancel");
            return true;
    }

}
```

（3）编写 TCC 控制类。

创建类 InventoryTccController，代码如下。

```
package com.cpf.inventory.controller;

import com.cpf.inventory.service.InventoryInterface;
import com.cpf.inventory.service.InventoryService;
import io.seata.spring.annotation.GlobalTransactional;
import org.springframework.beans.factory.annotation.Autowired;
import org.springframework.web.bind.annotation.GetMapping;
import org.springframework.web.bind.annotation.RestController;

/**
 * Seata AT 模式 测试控制层
 * @author 晁鹏飞
 */
@RestController
public class InventoryTccController {


    @Autowired
    private InventoryInterface inventoryInterface;


    @GetMapping("/inventory-tcc")
    @GlobalTransactional(rollbackFor = Exception.class)
    public String twoTcc(){

        inventoryInterface.inventoryTry(null);
        //  int i = 1/0;
        return "success";
    }
}
```

10.13.3 测试

下面分别介绍如下两种测试。

1. try 和 confirm 测试

为了方便 TCC 的测试，将通过观察控制台来进行验证。首先通过 Postman 使用 get 方法请求地址 localhost:2001/order-tcc，然后查看日志结果。

订单服务日志如下。

```
2021-04-11 17:36:06.188  INFO 10212 --- [nio-2001-exec-1] i.seata.tm.api.
DefaultGlobalTransaction  : Begin new global transaction
[192.168.56.1:8091:2071852823]
order try
2021-04-11 17:36:06.442  INFO 10212 --- [tch_RMROLE_1_16] i.s.core.rpc.
netty.RmMessageListener    : onMessage:xid=192.168.56.1:8091:2071852823,
branchId=2071852825,branchType=TCC,resourceId=orderAction,applicationData=
{"actionContext":{"sys::rollback":"orderRollback","sys::commit":
"orderCommit","action-start-time":1618133766191,"host-name":
"192.168.56.1","sys::prepare":"orderTry","actionName":"orderAction"}}
2021-04-11 17:36:06.443  INFO 10212 --- [tch_RMROLE_1_16] io.seata.rm.
AbstractRMHandler         : Branch committing: 192.168.56.1:8091:
2071852823 2071852825 orderAction {"actionContext":{"sys::rollback":
"orderRollback","sys::commit":"orderCommit","action-start-time":
1618133766191,"host-name":"192.168.56.1","sys::prepare":"orderTry",
"actionName":"orderAction"}}
order confirm
2021-04-11 17:36:06.458  INFO 10212 --- [tch_RMROLE_1_16] io.seata.rm.
AbstractResourceManager    : TCC resource commit result :true, xid:192.
168.56.1:8091:2071852823, branchId:2071852825, resourceId:orderAction
2021-04-11 17:36:06.459  INFO 10212 --- [tch_RMROLE_1_16] io.seata.rm.
AbstractRMHandler         : Branch commit result: PhaseTwo_Committed
2021-04-11 17:36:06.505  INFO 10212 --- [nio-2001-exec-1] i.seata.tm.api.
DefaultGlobalTransaction  : [192.168.56.1:8091:2071852823] commit status:
Committed
```

查看日志中的关键词 order try 和 order confirm，发现执行了 try 和 confirm 测试。

库存服务日志如下。

```
2021-04-11 17:36:06.354  INFO 7028 --- [nio-2002-exec-5] io.seata.tm.
TransactionManagerHolder    : TransactionManager Singleton io.seata.tm.
DefaultTransactionManager@2159e8cf
inventory try
2021-04-11 17:36:06.472  INFO 7028 --- [tch_RMROLE_1_16] i.s.core.rpc.netty.
RmMessageListener     : onMessage:xid=192.168.56.1:8091:2071852823,
branchId=2071852828,branchType=TCC,resourceId=inventoryTccAction,
applicationData={"actionContext":{"sys::rollback":"inventoryRollback",
"sys::commit":"inventoryCommit","action-start-time":1618133766358,
"host-name":"192.168.56.1","sys::prepare":"inventoryTry","actionName":
"inventoryTccAction"}}
2021-04-11 17:36:06.472  INFO 7028 --- [tch_RMROLE_1_16] io.seata.rm.
AbstractRMHandler        : Branch committing: 192.168.56.1:8091:
2071852823
2071852828 inventoryTccAction {"actionContext":{"sys::rollback":
"inventoryRollback","sys::commit":"inventoryCommit","action-start-time":
1618133766358,"host-name":"192.168.56.1","sys::prepare":"inventoryTry",
"actionName":"inventoryTccAction"}}
inventory confirm
2021-04-11 17:36:06.486  INFO 7028 --- [tch_RMROLE_1_16] io.seata.rm.
AbstractResourceManager     : TCC resource commit result :true, xid:192.
168.56.1:8091:2071852823, branchId:2071852828, resourceId:inventoryTccAction
2021-04-11 17:36:06.487  INFO 7028 --- [tch_RMROLE_1_16] io.seata.rm.
AbstractRMHandler        : Branch commit result: PhaseTwo_Committed
2021-04-11 17:39:26.366  INFO 7028 --- [trap-executor-0] c.n.d.s.r.aws.
ConfigClusterResolver     : Resolving eureka endpoints via configuration
```

查看日志中的关键词 inventory try 和 inventory confirm，发现执行了 try 和 confirm 测试。

2. try 和 cancel 测试

在程序中加入异常。例如，在订单服务 TCC 的实现类 OrderInterfaceImpl 中添加异常。

```
/**
 * Try 方法
 * @param businessActionContext
 * @return
```

```java
*/
@Override
@Transactional
public String orderTry(BusinessActionContext businessActionContext) {
    System.out.println("order try");

    inventoryTcc();
    System.out.println(1/0);
    return null;
}
```

下面再来进行测试。

首先通过 Postman 使用 get 方法请求地址 localhost:2001/order-tcc，然后查看日志结果。订单服务日志如下。

```
2021-04-11 17:48:25.727  INFO 23396 --- [nio-2001-exec-1] i.s.common.loader.EnhancedServiceLoader  : load LoadBalance[null] extension by class[io.seata.discovery.loadbalance.RandomLoadBalance]
2021-04-11 17:48:25.738  INFO 23396 --- [nio-2001-exec-1] i.seata.tm.api.DefaultGlobalTransaction  : Begin new global transaction [192.168.56.1:8091:2071852840]
order try
2021-04-11 17:48:26.075  INFO 23396 --- [tch_RMROLE_1_16] i.s.core.rpc.netty.RmMessageListener  : onMessage:xid=192.168.56.1:8091:2071852840,branchId=2071852842,branchType=TCC,resourceId=orderTry,applicationData={"actionContext":{"sys::rollback":"orderRollback","sys::commit":"orderCommit","action-start-time":1618134505741,"host-name":"192.168.56.1","sys::prepare":"orderTry","actionName":"orderTry"}}
2021-04-11 17:48:26.077  INFO 23396 --- [tch_RMROLE_1_16] io.seata.rm.AbstractRMHandler  : Branch Rollbacking: 192.168.56.1:8091:2071852840 2071852842 orderTry
order cancel
2021-04-11 17:48:26.091  INFO 23396 --- [tch_RMROLE_1_16] io.seata.rm.AbstractResourceManager  : TCC resource rollback result :true, xid:192.168.56.1:8091:2071852840, branchId:2071852842, resourceId:orderTry
2021-04-11 17:48:26.091  INFO 23396 --- [tch_RMROLE_1_16] io.seata.rm.AbstractRMHandler  : Branch Rollbacked result: PhaseTwo_Rollbacked
```

```
2021-04-11 17:48:26.116  INFO 23396 --- [nio-2001-exec-1] i.seata.tm.api.
DefaultGlobalTransaction   : [192.168.56.1:8091:2071852840] rollback status:
Rollbacked
2021-04-11 17:48:26.125 ERROR 23396 --- [nio-2001-exec-1] o.a.c.c.C.[.[.[/].
[dispatcherServlet]      : Servlet.service() for servlet [dispatcherServlet]
in context with path [] threw exception [Request processing failed; nested
exception is java.lang.ArithmeticException: / by zero] with root cause

java.lang.ArithmeticException: / by zero
    at com.cpf.order.service.impl.OrderInterfaceImpl.orderTry
(OrderInterfaceImpl.java:28) ~[classes/:na]
```

查看日志中的关键词 order try 和 order cancel，发现执行了 try 和 cancel 测试。

库存服务日志如下。

```
2021-04-11 17:48:25.988  INFO 23964 --- [nio-2002-exec-1] i.s.common.loader.
EnhancedServiceLoader   : load LoadBalance[null] extension by class[io.seata.
discovery.loadbalance.RandomLoadBalance]
inventory try
2021-04-11 17:48:26.043  INFO 23964 --- [tch_RMROLE_1_16] i.s.core.rpc.
netty.RmMessageListener       : onMessage:xid=192.168.56.1:8091:2071852840,
branchId=2071852844,branchType=TCC,resourceId=inventoryTry,applicationData=
{"actionContext":{"sys::rollback":"inventoryRollback","sys::commit":
"inventoryCommit","action-start-time":1618134505951,"host-name":
"192.168.56.1","sys::prepare":"inventoryTry","actionName":"inventoryTry"}}
2021-04-11 17:48:26.044  INFO 23964 --- [tch_RMROLE_1_16] io.seata.rm.
AbstractRMHandler            : Branch Rollbacking: 192.168.56.1:8091:
2071852840 2071852844 inventoryTry
inventory cancel
2021-04-11 17:48:26.059  INFO 23964 --- [tch_RMROLE_1_16] io.seata.rm.
AbstractResourceManager : TCC resource rollback result :true, xid:
192.168.56.1:8091:2071852840, branchId:2071852844, resourceId:inventoryTry
2021-04-11 17:48:26.059  INFO 23964 --- [tch_RMROLE_1_16] io.seata.rm.
AbstractRMHandler            : Branch Rollbacked result: PhaseTwo_Rollbacked
```

查看日志中的关键词 inventory try 和 inventory cancel，发现执行了 try 和 cancel 测试。

10.14 最大努力通知方案

10.14.1 什么是最大努力通知方案

如果接入过支付宝或者微信的支付接口，会遇到这样一种流程。例如，App 调用支付宝或微信的 SDK（software development kit，软件开发工具包）进行了支付，钱已经从用户的支付宝或微信账户转到了公司（开发 App 的公司）的支付宝或微信账户上，但是支付系统并不知道钱是否已经支付成功，需要支付宝或者微信回调公司的支付系统，才能进行后续的业务，流程如图 10-25 所示。

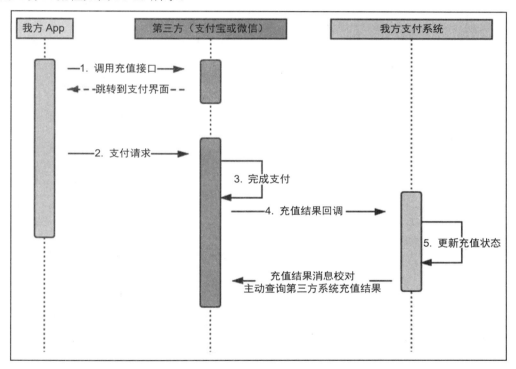

图 10-25　最大努力通知的支付解决方案流程

这其实就是一个最大努力通知的解决方案。在方案中需要保证以下两点。

☑ 消息重复通知机制：因为接收通知方（图 10-25 中的我方支付系统）可能没有接

收到通知，此时要有一定的机制对消息进行重复通知。
- ☑ 消息校对机制：如果尽最大努力也没有通知到接收方，或者接收方消费消息后要再次消费，此时可由接收方主动向通知方查询消息信息来满足需求。

其实这种解决方案针对内部系统和外部系统有不同的做法。
- ☑ 公司内部系统：针对公司内部系统来做的话，可以通过系统直接订阅消息队列来完成。因为都是自己的系统，直接订阅就可以。
- ☑ 公司外部系统：针对公司外部系统来做的话，直接让消费方订阅消息队列就有点不合适了，毕竟不能让两家公司同时对一个消息队列进行操作，所以，此时可以在内部写一个程序来订阅消息队列，通过 RPC 的方式调用消费方，使其被动地接收通知消息。在接支付宝和微信时，一般都是采用这种方式。

10.14.2 最大努力通知方案实战

业务场景如下：系统提前创建好订单（订单是未支付状态），然后调用第三方支付系统（写程序来模拟）进行支付，支付成功后，第三方系统通知我方系统进行订单的更新，将我方系统中的订单更改成已支付状态。

先针对内部系统进行实战，即让消费者直接订阅消息队列。

首先准备以下两个项目。
- ☑ third-party-pay：模拟第三方支付系统。
- ☑ my-pay：我方支付系统，用于接收第三方系统的回调。

1. third-party-pay 系统的搭建及开发

third-party-pay 系统的搭建及开发步骤如下。

（1）准备数据库。

创建数据库 notify-third-party-pay，在该数据库中新建两张表：tbl_order_pay 和 transaction_log。

表 tbl_order_pay 是第三方的支付记录表，建表语句如下。

```
CREATE TABLE `tbl_order_pay` (
  `id` int(16) NOT NULL AUTO_INCREMENT COMMENT 'id',
  `order_id` int(16) DEFAULT NULL COMMENT '订单Id',
  `order_status` int(2) DEFAULT NULL COMMENT '订单支付状态: 1：未支付, 2：已支付',
  PRIMARY KEY (`id`)
) ENGINE=InnoDB AUTO_INCREMENT=9 DEFAULT CHARSET=utf8;
```

表 transaction_log 是交易记录表，建表语句如下。

```
CREATE TABLE `transaction_log` (
  `id` varchar(32) CHARACTER SET utf8mb4 COLLATE utf8mb4_bin NOT NULL COMMENT '事务ID',
  `business` varchar(32) CHARACTER SET utf8mb4 COLLATE utf8mb4_bin NOT NULL COMMENT '业务标识',
  `foreign_key` varchar(32) CHARACTER SET utf8mb4 COLLATE utf8mb4_bin NOT NULL COMMENT '对应业务表中的主键',
  PRIMARY KEY (`id`)
) ENGINE=InnoDB DEFAULT CHARSET=utf8mb4 COLLATE=utf8mb4_bin;
```

（2）创建一个 Spring Boot 的 Web 项目 third-party-pay。

项目结构如图 10-26 所示。

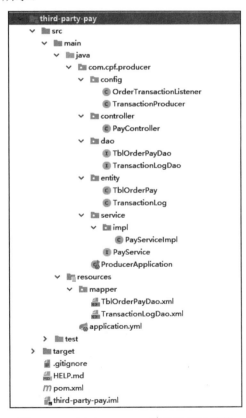

图 10-26　third-party-pay 项目的目录结构

（3）修改 pom 文件。

```xml
<?xml version="1.0" encoding="UTF-8"?>
<project xmlns="http://maven.apache.org/POM/4.0.0" xmlns:xsi="http://www.w3.org/2001/XMLSchema-instance"
    xsi:schemaLocation="http://maven.apache.org/POM/4.0.0 https://maven.apache.org/xsd/maven-4.0.0.xsd">
    <modelVersion>4.0.0</modelVersion>
    <parent>
        <groupId>org.springframework.boot</groupId>
        <artifactId>spring-boot-starter-parent</artifactId>
        <version>2.3.3.RELEASE</version>
        <relativePath/> <!-- lookup parent from repository -->
    </parent>
    <groupId>com.cpf.producer</groupId>
    <artifactId>third-party-pay</artifactId>
    <version>0.0.1-SNAPSHOT</version>
    <name>third-party-pay</name>
    <description>Demo project for Spring Boot</description>

    <properties>
        <java.version>1.8</java.version>
    </properties>

    <dependencies>
        <!--rocketmq-->
        <dependency>
            <groupId>org.apache.rocketmq</groupId>
            <artifactId>rocketmq-spring-boot-starter</artifactId>
            <version>2.0.2</version>
        </dependency>

        <!-- json-lib -->
        <dependency>
            <groupId>net.sf.json-lib</groupId>
            <artifactId>json-lib</artifactId>
            <version>2.4</version>
            <classifier>jdk15</classifier>
```

```xml
</dependency>

<dependency>
    <groupId>org.springframework.boot</groupId>
    <artifactId>spring-boot-starter-web</artifactId>
</dependency>

<dependency>
    <groupId>org.springframework.boot</groupId>
    <artifactId>spring-boot-starter-test</artifactId>
    <scope>test</scope>
    <exclusions>
        <exclusion>
            <groupId>org.junit.vintage</groupId>
            <artifactId>junit-vintage-engine</artifactId>
        </exclusion>
    </exclusions>
</dependency>

<dependency>
    <groupId>org.springframework</groupId>
    <artifactId>spring-tx</artifactId>
    <version>5.2.8.RELEASE</version>
</dependency>

<dependency>
    <groupId>org.springframework.boot</groupId>
    <artifactId>spring-boot-starter-web</artifactId>
</dependency>

<!-- MyBatis 相关依赖 -->
<dependency>
    <groupId>org.mybatis.spring.boot</groupId>
    <artifactId>mybatis-spring-boot-starter</artifactId>
    <version>2.0.0</version>
</dependency>
```

```xml
<!-- MySQL 驱动 -->
<dependency>
    <groupId>mysql</groupId>
    <artifactId>mysql-connector-java</artifactId>
</dependency>

<!-- 阿里巴巴数据库连接池 -->
<dependency>
    <groupId>com.alibaba</groupId>
    <artifactId>druid</artifactId>
    <version>1.1.12</version>
</dependency>

<dependency>
    <groupId>org.projectlombok</groupId>
    <artifactId>lombok</artifactId>
    <optional>true</optional>
</dependency>


<dependency>
    <groupId>org.springframework.boot</groupId>
    <artifactId>spring-boot-starter</artifactId>
</dependency>

<dependency>
    <groupId>org.springframework.boot</groupId>
    <artifactId>spring-boot-starter-test</artifactId>
    <scope>test</scope>
    <exclusions>
        <exclusion>
            <groupId>org.junit.vintage</groupId>
            <artifactId>junit-vintage-engine</artifactId>
        </exclusion>
    </exclusions>
</dependency>
</dependencies>
```

```xml
<build>
    <plugins>
        <plugin>
            <groupId>org.springframework.boot</groupId>
            <artifactId>spring-boot-maven-plugin</artifactId>
        </plugin>
    </plugins>
</build>
</project>
```

（4）修改 application.yml。

```yaml
server:
  port: 8081

spring:
  application:
    name: notify-third-party-pay
  datasource:
    type: com.alibaba.druid.pool.DruidDataSource
    driver-class-name: com.mysql.cj.jdbc.Driver
    url: jdbc:mysql://localhost:3306/notify-third-party-pay?characterEncoding=UTF-8&serverTimezone=Asia/Shanghai
    username: root
    password: root
    dbcp2:
      initial-size: 5
      min-idle: 5
      max-total: 5
      max-wait-millis: 200
      validation-query: SELECT 1
      test-while-idle: true
      test-on-borrow: false
      test-on-return: false

mybatis:
  mapper-locations:
    - classpath:mapper/*.xml
```

（5）编写 mapper 文件。

在 resources 目录下创建目录 mapper。编写数据库访问的 mapper 文件，即表 tbl_order_pay 的操作文件 TblOrderPayDao.xml。

```xml
<?xml version="1.0" encoding="UTF-8"?>
<!DOCTYPE mapper PUBLIC "-//mybatis.org//DTD Mapper 3.0//EN" "http://mybatis.org/dtd/mybatis-3-mapper.dtd">
<mapper namespace="com.cpf.producer.dao.TblOrderPayDao">
  <resultMap id="BaseResultMap" type="com.cpf.producer.entity.TblOrderPay">
    <id column="id" jdbcType="INTEGER" property="id" />
    <result column="order_id" jdbcType="INTEGER" property="orderId" />
    <result column="order_status" jdbcType="INTEGER" property="orderStatus" />
  </resultMap>
  <sql id="Base_Column_List">
    id, order_id, order_status
  </sql>
  <select id="selectByPrimaryKey" parameterType="java.lang.Integer" resultMap="BaseResultMap">
    select
    <include refid="Base_Column_List" />
    from tbl_order_pay
    where id = #{id,jdbcType=INTEGER}
  </select>
  <select id="selectByOrderId" parameterType="java.lang.Integer" resultMap="BaseResultMap">
    select
    <include refid="Base_Column_List" />
    from tbl_order_pay
    where order_id = #{orderId,jdbcType=INTEGER}
  </select>
  <delete id="deleteByPrimaryKey" parameterType="java.lang.Integer">
    delete from tbl_order_pay
    where id = #{id,jdbcType=INTEGER}
  </delete>
  <insert id="insert" keyColumn="id" keyProperty="id" parameterType="com.cpf.producer.entity.TblOrderPay" useGeneratedKeys="true">
```

```xml
      insert into tbl_order_pay (order_id, order_status)
      values (#{orderId,jdbcType=INTEGER}, #{orderStatus,jdbcType=INTEGER})
    </insert>
    <insert id="insertSelective" keyColumn="id" keyProperty="id"
parameterType="com.cpf.producer.entity.TblOrderPay" useGeneratedKeys="true">
      insert into tbl_order_pay
      <trim prefix="(" suffix=")" suffixOverrides=",">
        <if test="orderId != null">
          order_id,
        </if>
        <if test="orderStatus != null">
          order_status,
        </if>
      </trim>
      <trim prefix="values (" suffix=")" suffixOverrides=",">
        <if test="orderId != null">
          #{orderId,jdbcType=INTEGER},
        </if>
        <if test="orderStatus != null">
          #{orderStatus,jdbcType=INTEGER},
        </if>
      </trim>
    </insert>
    <update id="updateByPrimaryKeySelective" parameterType="com.cpf.producer.entity.TblOrderPay">
      update tbl_order_pay
      <set>
        <if test="orderId != null">
          order_id = #{orderId,jdbcType=INTEGER},
        </if>
        <if test="orderStatus != null">
          order_status = #{orderStatus,jdbcType=INTEGER},
        </if>
      </set>
      where id = #{id,jdbcType=INTEGER}
    </update>
    <update id="updateByPrimaryKey" parameterType="com.cpf.producer.entity.
```

```xml
TblOrderPay">
    update tbl_order_pay
    set order_id = #{orderId,jdbcType=INTEGER},
      order_status = #{orderStatus,jdbcType=INTEGER}
    where id = #{id,jdbcType=INTEGER}
  </update>
</mapper>
```

表 transaction_log 的操作文件 TransactionLogDao.xml 如下。

```xml
<?xml version="1.0" encoding="UTF-8"?>
<!DOCTYPE mapper PUBLIC "-//mybatis.org//DTD Mapper 3.0//EN" "http://mybatis.org/dtd/mybatis-3-mapper.dtd">
<mapper namespace="com.cpf.producer.dao.TransactionLogDao">
  <resultMap id="BaseResultMap" type="com.cpf.producer.entity.TransactionLog">
    <id column="id" jdbcType="VARCHAR" property="id" />
    <result column="business" jdbcType="VARCHAR" property="business" />
    <result column="foreign_key" jdbcType="VARCHAR" property="foreignKey" />
  </resultMap>
  <sql id="Base_Column_List">
    id, business, foreign_key
  </sql>
  <select id="selectByPrimaryKey" parameterType="java.lang.String" resultMap="BaseResultMap">
    select
    <include refid="Base_Column_List" />
    from transaction_log
    where id = #{id,jdbcType=VARCHAR}
  </select>
  <delete id="deleteByPrimaryKey" parameterType="java.lang.String">
    delete from transaction_log
    where id = #{id,jdbcType=VARCHAR}
  </delete>
  <insert id="insert" keyColumn="id" keyProperty="id" parameterType="com.cpf.producer.entity.TransactionLog" useGeneratedKeys="true">
    insert into transaction_log (id,business, foreign_key)
    values (#{id,jdbcType=VARCHAR},#{business,jdbcType=VARCHAR},
```

```xml
        #{foreignKey,jdbcType=VARCHAR})
  </insert>
  <insert id="insertSelective" keyColumn="id" keyProperty="id" parameterType="com.cpf.producer.entity.TransactionLog" useGeneratedKeys="true">
    insert into transaction_log
    <trim prefix="(" suffix=")" suffixOverrides=",">
      <if test="business != null">
        business,
      </if>
      <if test="foreignKey != null">
        foreign_key,
      </if>
    </trim>
    <trim prefix="values (" suffix=")" suffixOverrides=",">
      <if test="business != null">
        #{business,jdbcType=VARCHAR},
      </if>
      <if test="foreignKey != null">
        #{foreignKey,jdbcType=VARCHAR},
      </if>
    </trim>
  </insert>
  <update id="updateByPrimaryKeySelective" parameterType="com.cpf.producer.entity.TransactionLog">
    update transaction_log
    <set>
      <if test="business != null">
        business = #{business,jdbcType=VARCHAR},
      </if>
      <if test="foreignKey != null">
        foreign_key = #{foreignKey,jdbcType=VARCHAR},
      </if>
    </set>
    where id = #{id,jdbcType=VARCHAR}
  </update>
  <update id="updateByPrimaryKey" parameterType="com.cpf.producer.entity.TransactionLog">
```

```xml
  update transaction_log
  set business = #{business,jdbcType=VARCHAR},
    foreign_key = #{foreignKey,jdbcType=VARCHAR}
  where id = #{id,jdbcType=VARCHAR}
</update>

<select id="selectCount" parameterType="java.lang.String" resultType="java.lang.Integer">
  select
  count(*)
  from transaction_log
  where id = #{id,jdbcType=VARCHAR}
</select>
</mapper>
```

（6）编写实体类。

在 com.cpf.producer.entity 包中创建实体类 TransactionLog.java，代码如下。

```java
package com.cpf.producer.entity;

import lombok.Data;

import java.io.Serializable;

/**
 * transaction_log
 * @author 晁鹏飞
 */
@Data
public class TransactionLog implements Serializable {
    /**
     * 事务 ID
     */
    private String id;

    /**
     * 订单业务号
     */
```

```
    private String business;

    /**
     * 对应业务表中的主键
     */
    private String foreignKey;

}
```

创建实体类 TblOrderPay.java，代码如下。

```
package com.cpf.producer.entity;

import java.io.Serializable;
import lombok.Data;

/**
 * tbl_order_pay
 * @author
 */
@Data
public class TblOrderPay implements Serializable {
    /**
     * id
     */
    private Integer id;

    /**
     * 订单Id
     */
    private Integer orderId;

    /**
     * 订单支付状态：1：未支付，2：已支付
     */
    private Integer orderStatus;

    private static final long serialVersionUID = 1L;
}
```

（7）编写对应的 DAO 操作类。

在 com.cpf.producer.dao 包中创建类 TblOrderPayDao.java，代码如下。

```java
package com.cpf.producer.dao;

import com.cpf.producer.entity.TblOrderPay;
import org.apache.ibatis.annotations.Mapper;

/**
 * TblOrderPay 数据操作层
 * @author
 */
@Mapper
public interface TblOrderPayDao {

    int deleteByPrimaryKey(Integer id);

    int insert(TblOrderPay record);

    int insertSelective(TblOrderPay record);

    TblOrderPay selectByPrimaryKey(Integer id);

    int updateByPrimaryKeySelective(TblOrderPay record);

    int updateByPrimaryKey(TblOrderPay record);

    TblOrderPay selectByOrderId(Integer orderId);
}
```

创建类 TransactionLogDao.java，代码如下。

```java
package com.cpf.producer.dao;

import com.cpf.producer.entity.TransactionLog;
import org.apache.ibatis.annotations.Mapper;

/**
 * TransactionLog 数据操作层
```

```
 * @author 晁鹏飞
 */
@Mapper
public interface TransactionLogDao {

    int deleteByPrimaryKey(String id);

    int insert(TransactionLog record);

    int insertSelective(TransactionLog record);

    TransactionLog selectByPrimaryKey(String id);

    int updateByPrimaryKeySelective(TransactionLog record);

    int updateByPrimaryKey(TransactionLog record);

    int selectCount(String id);
}
```

（8）编写 Service 层。

创建包 com.cpf.producer.service，定义接口 PayService.java，代码如下。

```
package com.cpf.producer.service;

import com.cpf.producer.entity.TblOrderPay;
import org.apache.rocketmq.client.exception.MQClientException;

/**
 * @author 马士兵教育:chaopengfei
 * @date 2022/3/12
 */
public interface PayService {

    public void pay(TblOrderPay tblOrderPay,String topic) throws MQClientException;

    public int insert(TblOrderPay tblOrderPay , String transactionId);
```

```
    public TblOrderPay query(Integer orderId);
}
```

定义实现类 PayServiceImpl.java，代码如下。

```java
package com.cpf.producer.service.impl;

import com.alibaba.fastjson.JSON;
import com.cpf.producer.config.TransactionProducer;
import com.cpf.producer.dao.TblOrderPayDao;
import com.cpf.producer.dao.TransactionLogDao;
import com.cpf.producer.entity.TblOrderPay;
import com.cpf.producer.entity.TransactionLog;
import com.cpf.producer.service.PayService;
import org.apache.rocketmq.client.exception.MQClientException;
import org.springframework.beans.factory.annotation.Autowired;
import org.springframework.stereotype.Service;
import org.springframework.transaction.annotation.Transactional;

/**
 * @author 马士兵教育:chaopengfei
 * @date 2022/3/12
 */
@Service
public class PayServiceImpl implements PayService {

    @Autowired
    TblOrderPayDao tblOrderPayDao;

    @Autowired
    TransactionLogDao transactionLogDao;

    @Autowired
    TransactionProducer transactionProducer;

    @Override
    public void pay(TblOrderPay tblOrderPay,String topic) throws MQClientException {
```

```java
        transactionProducer.send(JSON.toJSONString(tblOrderPay),topic);
    }

    @Transactional
    @Override
    public int insert(TblOrderPay tblOrderPay , String transactionId) {
        // 写入事务日志
        TransactionLog log = new TransactionLog();
        log.setId(transactionId);
        log.setBusiness(tblOrderPay.getOrderId()+"");
        log.setForeignKey(tblOrderPay.getId()+"");
        transactionLogDao.insert(log);

        return tblOrderPayDao.insertSelective(tblOrderPay);
    }

    @Override
    public TblOrderPay query(Integer orderId) {
        return tblOrderPayDao.selectByOrderId(orderId);
    }
}
```

（9）定义发送消息类。

创建包 com.cpf.producer.config，在包中创建类 OrderTransactionListener.java，定义事务消息监听器，代码如下。

```java
package com.cpf.producer.config;

import com.alibaba.fastjson.JSONObject;
import com.cpf.producer.dao.TransactionLogDao;
import com.cpf.producer.entity.TblOrderPay;
import com.cpf.producer.service.PayService;
import lombok.extern.slf4j.Slf4j;
import org.apache.rocketmq.client.producer.LocalTransactionState;
import org.apache.rocketmq.client.producer.TransactionListener;
import org.apache.rocketmq.common.message.Message;
import org.apache.rocketmq.common.message.MessageExt;
import org.springframework.beans.factory.annotation.Autowired;
```

```java
import org.springframework.stereotype.Component;

/**
 * 事务监听器
 * @author 晁鹏飞
 */
@Component
@Slf4j
public class OrderTransactionListener implements TransactionListener {

    @Autowired
    PayService payService;

    @Autowired
    TransactionLogDao transactionLogDao;

    /**
     *
     * @param message
     * @param o
     * @return
     */
    @Override
    public LocalTransactionState executeLocalTransaction(Message message, Object o) {
        log.info("开始执行本地事务...");
        LocalTransactionState state;
        String keys = message.getKeys();
        System.out.println("keys:"+keys);
        try{
            String body = new String(message.getBody());
            TblOrderPay order = JSONObject.parseObject(body, TblOrderPay.class);
            payService.insert(order,message.getTransactionId());

            state = LocalTransactionState.COMMIT_MESSAGE;
            // state = LocalTransactionState.UNKNOW;
            log.info("本地事务已提交。{}",message.getTransactionId());
```

```
        }catch (Exception e){
            log.info("执行本地事务失败。{}",e);
            state = LocalTransactionState.ROLLBACK_MESSAGE;
        }
        return state;
    }

    /**
     *
     * @param messageExt
     * @return
     */
    @Override
    public LocalTransactionState checkLocalTransaction(MessageExt messageExt) {

        // 如果回查发生多次失败,则发邮件提醒,进行人工回查
        log.info("开始回查本地事务状态。{}",messageExt.getTransactionId());
        LocalTransactionState state;
        String transactionId = messageExt.getTransactionId();
        if (transactionLogDao.selectCount(transactionId)>0){
            state = LocalTransactionState.COMMIT_MESSAGE;
        }else {
            state = LocalTransactionState.ROLLBACK_MESSAGE;
        }
        System.out.println();
        log.info("结束本地事务状态查询:{}",state);
        return state;
    }
}
```

这个类执行了事务消息中本地数据库操作的业务逻辑。

定义消息发送组件,创建类 TransactionProducer.java,代码如下。

```
package com.cpf.producer.config;

import org.apache.rocketmq.client.exception.MQClientException;
import org.apache.rocketmq.client.producer.TransactionMQProducer;
```

```java
import org.apache.rocketmq.client.producer.TransactionSendResult;
import org.apache.rocketmq.common.message.Message;
import org.springframework.beans.factory.annotation.Autowired;
import org.springframework.stereotype.Component;

import javax.annotation.PostConstruct;
import java.util.concurrent.ArrayBlockingQueue;
import java.util.concurrent.ThreadPoolExecutor;
import java.util.concurrent.TimeUnit;

/**
 * 消息发送方工具类
 * @author 晁鹏飞
 */
@Component
public class TransactionProducer {

    private String producerGroup = "order_trans_group";

    // 事务消息
    private TransactionMQProducer producer;

    /**
     * 用于执行本地事务和事务状态回查的监听器
     */
    @Autowired
    OrderTransactionListener orderTransactionListener;

    // 执行任务的线程池
    ThreadPoolExecutor executor = new ThreadPoolExecutor(5, 10, 60,
            TimeUnit.SECONDS, new ArrayBlockingQueue<>(50));

    @PostConstruct
    public void init(){
        producer = new TransactionMQProducer(producerGroup);
        producer.setNamesrvAddr("127.0.0.1:9876");
        producer.setSendMsgTimeout(Integer.MAX_VALUE);
```

```java
        producer.setExecutorService(executor);
        producer.setTransactionListener(orderTransactionListener);
        this.start();
    }

    private void start(){
        try {
            this.producer.start();
        } catch (MQClientException e) {
            e.printStackTrace();
        }
    }

    /**
     * 事务消息发送
     * @param data
     * @param topic
     * @return
     * @throws MQClientException
     */
    public TransactionSendResult send(String data, String topic) throws
MQClientException {
        Message message = new Message(topic,data.getBytes());
        return this.producer.sendMessageInTransaction(message, null);
    }
}
```

（10）定义 Controller，即用户支付的入口。

创建包 com.cpf.producer.controller，在包中创建 Controller，即 PayController.java，代码如下。

```java
package com.cpf.producer.controller;

import com.cpf.producer.config.TransactionProducer;
import com.cpf.producer.entity.TblOrderPay;
import com.cpf.producer.service.PayService;
import lombok.extern.slf4j.Slf4j;
import org.apache.rocketmq.client.exception.MQClientException;
```

```java
import org.springframework.beans.factory.annotation.Autowired;
import org.springframework.web.bind.annotation.*;

/**
 * 事务消息Controller
 * @author 晁鹏飞
 */
@RestController
@Slf4j
public class PayController {

    @Autowired
    PayService payService;

    /**
     * 支付业务接口
     * @param orderPay
     * @throws MQClientException
     */
    @PostMapping("/pay")
    public void pay(@RequestBody TblOrderPay orderPay) throws MQClientException {
        payService.pay(orderPay,"pay-order");
    }

    /**
     * 支付结果查询接口
     * @param orderId
     * @return
     * @throws MQClientException
     */
    @GetMapping("/pay-query/{orderId}")
    public TblOrderPay payQuery(@PathVariable("orderId") Integer orderId) throws MQClientException {
        return payService.queryByOrderId(orderId);
    }
}
```

2. my-pay 系统的搭建及开发

my-pay 系统的搭建及开发步骤如下。

（1）创建订单数据库并新建表。

创建数据库 notify-my-pay，并新建表 tbl_my_order_pay，建表语句如下。

```
CREATE TABLE `tbl_my_order_pay` (
  `id` int(16) NOT NULL AUTO_INCREMENT COMMENT 'id',
  `order_id` int(16) DEFAULT NULL COMMENT '订单 Id',
  `order_status` int(2) DEFAULT NULL COMMENT '订单支付状态：1：未支付，2：已支付',
  PRIMARY KEY (`id`)
) ENGINE=InnoDB AUTO_INCREMENT=5 DEFAULT CHARSET=utf8;
```

（2）创建一个 Spring Boot 的 Web 项目 my-pay。

my-pay 项目的目录结构如图 10-27 所示。

图 10-27　my-pay 项目的目录结构

（3）修改 pom 文件，引入相关依赖。

```xml
<?xml version="1.0" encoding="UTF-8"?>
<project xmlns="http://maven.apache.org/POM/4.0.0" xmlns:xsi="http://www.w3.org/2001/XMLSchema-instance"
    xsi:schemaLocation="http://maven.apache.org/POM/4.0.0 https://maven.apache.org/xsd/maven-4.0.0.xsd">
    <modelVersion>4.0.0</modelVersion>
    <parent>
        <groupId>org.springframework.boot</groupId>
        <artifactId>spring-boot-starter-parent</artifactId>
        <version>2.3.3.RELEASE</version>
        <relativePath/> <!-- lookup parent from repository -->
    </parent>
    <groupId>com.cpf.consumer</groupId>
    <artifactId>my-pay</artifactId>
    <version>0.0.1-SNAPSHOT</version>
    <name>my-pay</name>
    <description>Demo project for Spring Boot</description>

    <properties>
        <java.version>1.8</java.version>
    </properties>

    <dependencies>
        <!--rocketmq-->
        <dependency>
            <groupId>org.apache.rocketmq</groupId>
            <artifactId>rocketmq-spring-boot-starter</artifactId>
            <version>2.0.2</version>
        </dependency>

        <!-- json-lib -->
        <dependency>
            <groupId>net.sf.json-lib</groupId>
            <artifactId>json-lib</artifactId>
            <version>2.4</version>
            <classifier>jdk15</classifier>
```

```xml
</dependency>


<dependency>
    <groupId>org.springframework.boot</groupId>
    <artifactId>spring-boot-starter-web</artifactId>
</dependency>

<dependency>
    <groupId>org.springframework.boot</groupId>
    <artifactId>spring-boot-starter-test</artifactId>
    <scope>test</scope>
    <exclusions>
        <exclusion>
            <groupId>org.junit.vintage</groupId>
            <artifactId>junit-vintage-engine</artifactId>
        </exclusion>
    </exclusions>
</dependency>


<dependency>
    <groupId>org.springframework</groupId>
    <artifactId>spring-tx</artifactId>
    <version>5.2.8.RELEASE</version>
</dependency>

<dependency>
    <groupId>org.springframework.boot</groupId>
    <artifactId>spring-boot-starter-web</artifactId>
</dependency>

<!-- MyBatis 相关依赖 -->
<dependency>
    <groupId>org.mybatis.spring.boot</groupId>
    <artifactId>mybatis-spring-boot-starter</artifactId>
    <version>2.0.0</version>
</dependency>
```

```xml
<!-- MySQL 驱动 -->
<dependency>
    <groupId>mysql</groupId>
    <artifactId>mysql-connector-java</artifactId>
</dependency>

<!-- 阿里巴巴数据库连接池 -->
<dependency>
    <groupId>com.alibaba</groupId>
    <artifactId>druid</artifactId>
    <version>1.1.12</version>
</dependency>

<dependency>
    <groupId>org.projectlombok</groupId>
    <artifactId>lombok</artifactId>
    <optional>true</optional>
</dependency>


<dependency>
    <groupId>org.springframework.boot</groupId>
    <artifactId>spring-boot-starter</artifactId>
</dependency>

<dependency>
    <groupId>org.springframework.boot</groupId>
    <artifactId>spring-boot-starter-test</artifactId>
    <scope>test</scope>
    <exclusions>
        <exclusion>
            <groupId>org.junit.vintage</groupId>
            <artifactId>junit-vintage-engine</artifactId>
        </exclusion>
    </exclusions>
</dependency>
```

```xml
        </dependencies>

    <build>
        <plugins>
            <plugin>
                <groupId>org.springframework.boot</groupId>
                <artifactId>spring-boot-maven-plugin</artifactId>
            </plugin>
        </plugins>
    </build>
</project>
```

（4）修改 application.yml。

```yaml
server:
  port: 8082

spring:
  application:
    name: my-pay
  datasource:
    type: com.alibaba.druid.pool.DruidDataSource
    driver-class-name: com.mysql.cj.jdbc.Driver
    url: jdbc:mysql://localhost:3306/notify-my-pay?characterEncoding=UTF-8&serverTimezone=Asia/Shanghai
    username: root
    password: root
    dbcp2:
      initial-size: 5
      min-idle: 5
      max-total: 5
      max-wait-millis: 200
      validation-query: SELECT 1
      test-while-idle: true
      test-on-borrow: false
      test-on-return: false

mybatis:
```

```yaml
  mapper-locations:
   - classpath:mapper/*.xml
logging:
  level:
    root: debug
```

注意，为了观察 SQL 预计的执行，使用了 debug 级别的日志。

（5）编写 mapper 文件。

在 resources 目录下创建目录 mapper，编写数据库访问的 mapper 文件，即表 tbl_my_order_pay 的操作文件 TblMyOrderPayDao.xml。

```xml
<?xml version="1.0" encoding="UTF-8"?>
<!DOCTYPE mapper PUBLIC "-//mybatis.org//DTD Mapper 3.0//EN" "http://mybatis.org/dtd/mybatis-3-mapper.dtd">
<mapper namespace="com.cpf.consumer.dao.TblMyOrderPayDao">
  <resultMap id="BaseResultMap" type="com.cpf.consumer.entity.TblMyOrderPay">
    <id column="id" jdbcType="INTEGER" property="id" />
    <result column="order_id" jdbcType="INTEGER" property="orderId" />
    <result column="order_status" jdbcType="INTEGER" property="orderStatus" />
  </resultMap>
  <sql id="Base_Column_List">
    id, order_id, order_status
  </sql>
  <select id="selectByPrimaryKey" parameterType="java.lang.Integer" resultMap="BaseResultMap">
    select
    <include refid="Base_Column_List" />
    from tbl_my_order_pay
    where id = #{id,jdbcType=INTEGER}
  </select>
  <delete id="deleteByPrimaryKey" parameterType="java.lang.Integer">
    delete from tbl_my_order_pay
    where id = #{id,jdbcType=INTEGER}
  </delete>
  <insert id="insert" keyColumn="id" keyProperty="id" parameterType="com.cpf.consumer.entity.TblMyOrderPay" useGeneratedKeys="true">
```

```xml
    insert into tbl_my_order_pay (order_id, order_status)
    values (#{orderId,jdbcType=INTEGER}, #{orderStatus,jdbcType=INTEGER})
  </insert>
  <insert id="insertSelective" keyColumn="id" keyProperty="id" parameterType="com.cpf.consumer.entity.TblMyOrderPay" useGeneratedKeys="true">
    insert into tbl_my_order_pay
    <trim prefix="(" suffix=")" suffixOverrides=",">
      <if test="orderId != null">
        order_id,
      </if>
      <if test="orderStatus != null">
        order_status,
      </if>
    </trim>
    <trim prefix="values (" suffix=")" suffixOverrides=",">
      <if test="orderId != null">
        #{orderId,jdbcType=INTEGER},
      </if>
      <if test="orderStatus != null">
        #{orderStatus,jdbcType=INTEGER},
      </if>
    </trim>
  </insert>
  <update id="updateByPrimaryKeySelective" parameterType="com.cpf.consumer.entity.TblMyOrderPay">
    update tbl_my_order_pay
    <set>
      <if test="orderId != null">
        order_id = #{orderId,jdbcType=INTEGER},
      </if>
      <if test="orderStatus != null">
        order_status = #{orderStatus,jdbcType=INTEGER},
      </if>
    </set>
    where id = #{id,jdbcType=INTEGER}
  </update>
```

```xml
<update id="updateByOrderId" parameterType="com.cpf.consumer.entity.TblMyOrderPay">
    update tbl_my_order_pay
    <set>
      <if test="orderId != null">
        order_id = #{orderId,jdbcType=INTEGER},
      </if>
      <if test="orderStatus != null">
        order_status = #{orderStatus,jdbcType=INTEGER},
      </if>
    </set>
    where order_id = #{orderId,jdbcType=INTEGER}
</update>

<update id="updateByPrimaryKey" parameterType="com.cpf.consumer.entity.TblMyOrderPay">
    update tbl_my_order_pay
    set order_id = #{orderId,jdbcType=INTEGER},
      order_status = #{orderStatus,jdbcType=INTEGER}
    where id = #{id,jdbcType=INTEGER}
</update>
</mapper>
```

（6）编写实体类。

在 **com.cpf.consumer.entity** 包中创建类 **TblMyOrderPay.java**，代码如下。

```java
package com.cpf.consumer.entity;

import java.io.Serializable;
import lombok.Data;

/**
 * tbl_my_order_pay
 * @author
 */
@Data
public class TblMyOrderPay implements Serializable {
    /**
```

```
 * id
 */
private Integer id;

/**
 * 订单Id
 */
private Integer orderId;

/**
 * 订单支付状态：此时的订单状态通过第三方系统的回调状态来定
 */
private Integer orderStatus;
}
```

(7) 编写对应的 DAO 操作类。

在 com.cpf.consumer.dao 包中创建类 TblMyOrderPayDao.java，代码如下。

```
package com.cpf.consumer.dao;

import com.cpf.consumer.entity.TblMyOrderPay;
import org.apache.ibatis.annotations.Mapper;

/**
 * TblMyOrderPay 数据操作层
 * @author 晁鹏飞
 */
@Mapper
public interface TblMyOrderPayDao {

    int deleteByPrimaryKey(Integer id);

    int insert(TblMyOrderPay record);

    int insertSelective(TblMyOrderPay record);

    TblMyOrderPay selectByPrimaryKey(Integer id);
```

```
    int updateByPrimaryKeySelective(TblMyOrderPay record);

    int updateByOrderId(TblMyOrderPay record);

    int updateByPrimaryKey(TblMyOrderPay record);
}
```

（8）编写 Service 层。

创建包 com.cpf.consumer.service，定义接口类 MyOrderPayService.java，代码如下。

```
package com.cpf.consumer.service;

import com.cpf.consumer.entity.TblMyOrderPay;

/**
 * MyOrderPay 接口层
 * @author 马士兵教育:chaopengfei
 * @date 2022/3/13
 */
public interface MyOrderPayService {

    public int update(TblMyOrderPay orderPay);
}
```

定义服务类 MyOrderPayServiceImpl.java，代码如下。

```
package com.cpf.consumer.service.impl;

import com.cpf.consumer.dao.TblMyOrderPayDao;
import com.cpf.consumer.entity.TblMyOrderPay;
import com.cpf.consumer.service.MyOrderPayService;
import org.springframework.beans.factory.annotation.Autowired;
import org.springframework.stereotype.Service;

/**
 * MyOrderPay 服务实现类
 * @author 马士兵教育:chaopengfei
 * @date 2022/3/13
```

```java
*/
@Service
public class MyOrderPayServiceImpl implements MyOrderPayService {

    @Autowired
    TblMyOrderPayDao myOrderPayDao;

    @Override
    public int update(TblMyOrderPay orderPay) {
        return myOrderPayDao.updateByOrderId(orderPay);
    }
}
```

（9）定义消息订阅组件。

创建包 com.cpf.consumer.compent，定义消费监听类 OrderListener.java，代码如下。

```java
package com.cpf.consumer.compent;

import com.alibaba.fastjson.JSONObject;
import com.cpf.consumer.dao.TblMyOrderPayDao;
import com.cpf.consumer.entity.TblMyOrderPay;
import com.cpf.consumer.service.MyOrderPayService;
import lombok.extern.slf4j.Slf4j;
import org.apache.rocketmq.client.consumer.listener.ConsumeConcurrentlyContext;
import org.apache.rocketmq.client.consumer.listener.ConsumeConcurrentlyStatus;
import org.apache.rocketmq.client.consumer.listener.MessageListenerConcurrently;
import org.apache.rocketmq.common.message.MessageExt;
import org.springframework.beans.factory.annotation.Autowired;
import org.springframework.stereotype.Component;

import java.util.List;

/**
 * 消息监听器
 * @author 晁鹏飞
```

```java
 */
@Component
@Slf4j
public class OrderListener implements MessageListenerConcurrently {

    @Autowired
    MyOrderPayService myOrderPayService;

    @Override
    public ConsumeConcurrentlyStatus consumeMessage(List<MessageExt> list, ConsumeConcurrentlyContext context) {
        log.info("消费者线程监听到消息。");
        try{

            // System.out.println(1/0);
            for (MessageExt message:list) {
                log.info("开始处理订单数据，准备修改订单状态...");
                TblMyOrderPay order = JSONObject.parseObject(message.getBody(), TblMyOrderPay.class);
                myOrderPayService.update(order);
            }
            return ConsumeConcurrentlyStatus.CONSUME_SUCCESS;
        }catch (Exception e){
            log.error("处理消费者数据发生异常。{}",e);
            return ConsumeConcurrentlyStatus.RECONSUME_LATER;
        }
    }
}
```

定义消费组件 Consumer，并对监听进行注册，代码如下。

```
package com.cpf.consumer.compent;

import org.apache.rocketmq.client.consumer.DefaultMQPushConsumer;
import org.apache.rocketmq.client.exception.MQClientException;
import org.springframework.beans.factory.annotation.Autowired;
import org.springframework.stereotype.Component;
```

```java
import javax.annotation.PostConstruct;

/**
 * 消费端定义
 * @author 晁鹏飞
 */
@Component
public class Consumer {

    String consumerGroup = "consumer-group";
    DefaultMQPushConsumer consumer;

    @Autowired
    OrderListener orderListener;

    @PostConstruct
    public void init() throws MQClientException {
        consumer = new DefaultMQPushConsumer(consumerGroup);
        consumer.setNamesrvAddr("127.0.0.1:9876");
        consumer.subscribe("pay-order","*");
        consumer.registerMessageListener(orderListener);

        // 两次失败，就进死信队列
        consumer.setMaxReconsumeTimes(2);
        consumer.start();
    }
}
```

目前消息消费者直接消费消息，就进行本地业务数据的操作。如果 my-pay 接入第三方，让第三方来回调 my-pay 接口，那么 my-pay 还需要提供一个 Controller，代码如下。

```java
package com.cpf.consumer.controller;

import com.cpf.consumer.entity.TblMyOrderPay;
import com.cpf.consumer.service.MyOrderPayService;
import org.springframework.beans.factory.annotation.Autowired;
import org.springframework.web.bind.annotation.*;
```

```java
/**
 * 支付结果回调控制类
 * @author 马士兵教育:chaopengfei
 * @date 2022/3/13
 */
@RestController
@RequestMapping
public class MyOrderPayController {

    @Autowired
    private MyOrderPayService myOrderPayService;

    /**
     * 支付成功回调接口
     * @param pay
     * @return
     */
    @PostMapping("/my-pay")
    public String myPay(@RequestBody TblMyOrderPay pay){

        myOrderPayService.update(pay);

        return "成功";
    }
}
```

10.15 小　　结

本章从分布式事务的业务场景出发，首先分析了分布式事务解决方案的思路，详细讲解了两阶段和三阶段提交协议，以及 CAP 定理和 BASE 理论在分布式事务中的应用。然后讲解了市面上主流的分布式事务解决方案：两阶段提交协议，三阶段提交协议，TCC 分步式事务解决方案、可靠消息最终一致性方案、最大努力通知解决方案。最后通过实践代码进行了几种分布式事务的落地，即 Seata 的 AT 模式和 TCC 模式，RocketMQ 事务消息方案，最大努力通知方案，达到让读者能在实际工作中应用的目的。

第 11 章 微服务鉴权认证安全设计

我们在平时使用互联网应用时，几乎都会用到授权的场景。例如，在用购物网站时，用户是否有权限查看已经购买过的商品列表；用户能否使用微博账号登录支付宝平台；用户能否使用云打印服务去读取网盘上的照片，并打印出来等。这就涉及客户端和服务间、服务与服务之间授权认证的问题。

在鉴权认证业务中主要有 3 种角色，以用户查看淘宝已购买商品列表为例，其角色如下。

- ☑ 资源拥有者。就是用户，拥有对商品列表的查看的权限。
- ☑ 客户应用。淘宝系统。
- ☑ 受保护的资源。就是商品列表。

11.1 鉴权认证常见的场景及解决方案

在生产环境中一般会有如下几种情况。

1. 将用户名和密码告知客户端应用

资源拥有者将用户名和密码告诉客户应用，这样客户应用就可以拿着用户名和密码，取得操作受保护资源的权限。这种场景适合于内部系统，如公司内部有两个系统，分别是审批系统和考勤系统，这时就可以用一个公司账号来登录，都可以查询员工的个人信息。由于账号、密码都是公司自己的，也不怕被非法的第三方知道。如果前面所说的两个系统都是公司外部的系统，此种情况就不适用了，因为公司的账号和密码会有被第三方窃取的风险。

2. 让客户应用和受保护的资源商定一个 key

这种情况适用于客户应用和受保护资源之间是授信关系的场景。例如，通过微信去访问大众点评，当决定用微信登录大众点评时，其实大众点评已经开发好利用微信登录的程序。大众点评是经过微信认证的，它和微信之间是授信关系。

3. 资源拥有者给客户应用一个临时密码，只能访问特定资源

这种情况类似于手机上应用的安全令牌，要临时登录某个系统时，需要输入临时令牌，这样在某个时间段内，就取得对某个资源的授权。

> **大厂面试**
>
> 面试官：能说一下你对认证和授权的认识吗？
> 回答分析：
> （1）Authentication（认证）是验证身份的凭据（如用户名/用户 ID 和密码），通过这个凭据，系统得以知道你就是你，也就是说系统存在你这个用户。所以，Authentication 被称为身份/用户验证。
> （2）Authorization（授权）发生在 Authentication 之后。授权，它主要掌管访问系统的权限。例如，有些特定资源只能是具有特定权限的用户才能访问，如 admin，有些对系统资源操作，如删除、添加、更新只能特定用户才具有。

11.1.1 单体应用

在单体应用中，一般的认证流程如下。

（1）客户端向服务端发送用户名和密码。

（2）服务端收到用户名和密码后进行校验，校验通过后，存一份 Session 在服务端，同时向用户的 Cookie 中下发一个 sessionId。

（3）下次用户需要访问时，就无须再输入用户名和密码，拿着 sessionId 就可以。

（4）服务端收到 sessionId 后，查询与之对应的 Session 信息就知道是哪个用户了，并执行相应的操作。

11.1.2 微服务应用

在微服务应用中，一般的认证流程如下。

（1）用户端（包括网页、无线原生的 App、服务器 Web 应用），向认证授权服务器发送用户名和密码。

（2）认证授权服务器进行用户名和密码的校验，校验通过后，向用户下发一个 Token。

（3）当用户端再访问系统中的其他服务时，就可以拿着 Token 直接访问了。

11.2　OAuth 2.0 介绍

在微服务架构下，或者在不同企业的服务之间发生调用时，就像 11.1 节提到的，如果将用户名和密码告知第三方客户应用，将存在很大的安全隐患。在主流的鉴权认证业务中，有一种方案 OAuth 2.0，它通过增加授信服务器的机制，解决了这个安全隐患问题。

OAuth 2.0 有如下几种角色。

（1）资源拥有者（resource owner，RO）。资源拥有者想要将自己的应用分发给客户应用。

（2）客户应用（client application，CA）。通常是一个 Web 应用或者 App 应用。

（3）授权服务器（authorization server，AS）。向资源拥有者申请授权，向客户应用下发访问令牌（Access Token）的服务器，并进行授权的管理。

（4）资源服务器（resource server，RS）。通常是一个站点或者服务 API，用于管理受保护的数据。

OAuth 2.0 整体流程如图 11-1 所示。

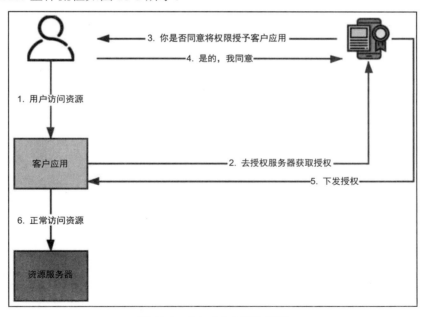

图 11-1　OAuth 2.0 整体流程

（1）用户通过客户应用访问资源。
（2）客户应用收到用户的请求后，向授权服务器获取授权。
（3）授权服务器再征得用户的同意。
（4）用户同意后，授权服务器向客户应用下发授权，也就是 Access Token。
（5）客户应用拿着 Access Token 访问资源。

其中的 Access Token 只赋予了客户应用有限的访问权限。例如，只能查看不能读写，只能 2min 内有效等。

这其实相当于生活中的一个场景，如笔者去阿里巴巴拜访老领导，当走到园区门口时，被保安拦住了，保安要求出示工牌，笔者说没有工牌并告诉保安是来拜访陈总的，保安要求去前台登记一下，当走到前台时，前台问笔者是否预约过，然后前台给预约的手机发一个验证码，笔者出示验证码后，前台发了一个临时工牌，然后笔者拿着临时工牌，保安就放行了。

如果在软件产品中用浏览器登录美团，可以用美团的账号和密码，也可以用微信登录。当用微信登录时，输入微信登录信息后，微信服务器会给美团发一个授权 Token，美团就可以拿着 Token 访问到微信昵称和头像信息。不用像注册美团新用户一样，重新设置昵称和头像信息了。

其实 OAuth 2.0 就是一种授权协议，保证第三方软件只有在获得授权后，才可以进一步访问授权者的数据。

OAuth 2.0 有如下优势。

- ☑ 比较安全，因为客户应用不需要接触用户的用户名和密码，授权服务器也更容易对权限进行集中保护。OAuth 2.0 本质上就是不让客户应用接触用户的用户名和密码。
- ☑ Access Token 是短寿命的，并且是可以携带用户信息的。
- ☑ 资源服务器和授权服务器进行解耦，便于权限的统一管理，不会侵入业务。
- ☑ 集中式授权，减少了客户应用开发的复杂度。
- ☑ 采用 HTTP/JSON 形式，易于请求和传递。
- ☑ 支持多种类型客户端的接入，如网页、原生 App 开发、微信等。

OAuth 2.0 的缺点如下。

- ☑ 协议框架太宽泛，造成各种实现的兼容性和互操作性特别差。
- ☑ OAuth 2.0 不是一个认证协议，它本身不会告诉你任何用户的信息。

OAuth 2.0 中的术语如下。

- ☑ 客户凭证：客户的用户名和密码。
- ☑ 令牌：由授权服务器颁发给第三方应用的令牌。
- ☑ 作用域：由资源拥有者对客户应用额外指定的细分权限。

OAuth 2.0 中令牌的类型如下。

- ☑ 访问令牌（access token）：用于代表一个用户或服务之间访问受保护的资源。
- ☑ 刷新令牌（refresh token）：用于去授权服务器获取一个新的访问令牌。
- ☑ 授权码（authorization code token）：用于交换获取访问令牌和刷新令牌。
- ☑ 持票令牌（bearer token）：不管谁拿到 Token 都可以访问资源，就像钞票一样，谁拿到都可以使用，安全性比较差。
- ☑ 校验占有权令牌（proof of possession token）：可以校验客户是否对 Token 有明确的拥有权。

其实，OAuth 2.0 只是一个框架，不支持 HTTP 以外的协议，不是一个认证协议，没有定义授权处理机制，没有定义 Token 的格式，没有定义加密方法，并不是一个单个的协议。它仅仅是授权框架，用于授权代理。

11.3 OAuth 2.0 实战

本节实践 OAuth 2.0 授权码许可模式（authorization code）。这是 OAuth 2.0 中最经典、最完备、最安全，也是应用最广泛的许可类型。下面新建 Spring Boot 项目 authcode-server，具体步骤如下。

（1）在 pom 文件中添加依赖。

```xml
<dependencies>
 <dependency>
   <groupId>org.springframework.boot</groupId>
   <artifactId>spring-boot-starter-security</artifactId>
 </dependency>
 <dependency>
   <groupId>org.springframework.boot</groupId>
   <artifactId>spring-boot-starter-web</artifactId>
 </dependency>
```

```xml
<!--for OAuth 2.0 -->
<dependency>
   <groupId>org.springframework.security.oauth</groupId>
   <artifactId>spring-security-oauth2</artifactId>
   <version>2.0.14.RELEASE</version>
</dependency>

<dependency>
   <groupId>org.springframework.boot</groupId>
   <artifactId>spring-boot-starter-test</artifactId>
   <scope>test</scope>
</dependency>
<dependency>
   <groupId>org.springframework.security</groupId>
   <artifactId>spring-security-test</artifactId>
   <scope>test</scope>
</dependency>
</dependencies>
```

（2）修改 application.yml 文件。

```yaml
security:
  user:
    name: cpf
    password: 123
```

（3）定义资源访问 API。

```java
package com.cpf.authcodeserver.api;

import org.springframework.http.ResponseEntity;
import org.springframework.security.core.context.SecurityContextHolder;
import org.springframework.security.core.userdetails.User;
import org.springframework.stereotype.Controller;
import org.springframework.web.bind.annotation.RequestMapping;

/**
 * 资源提供控制类
 */
```

```java
@Controller
public class UserController {

/**
 * 资源API,用于访问用户信息
 */
    @RequestMapping("/api/userinfo")
    public ResponseEntity<UserInfo> getUserInfo() {
        User user = (User) SecurityContextHolder.getContext()
                .getAuthentication().getPrincipal();
        String email = user.getUsername() + "@spring2go.com";

        UserInfo userInfo = new UserInfo();
        userInfo.setName(user.getUsername());
        userInfo.setEmail(email);

        return ResponseEntity.ok(userInfo);
    }

}
```

用户信息类 UserInfo 的代码如下。

```java
package com.cpf.authcodeserver.api;

public class UserInfo {

    private String name;

    private String email;

    public String getName() {
        return name;
    }

    public void setName(String name) {
        this.name = name;
    }
```

```java
    public String getEmail() {
        return email;
    }

    public void setEmail(String email) {
        this.email = email;
    }
}
```

（4）定义一个授权服务器。

```java
package com.cpf.authcodeserver.config;

import org.springframework.context.annotation.Configuration;
import org.springframework.security.oauth2.config.annotation.configurers.ClientDetailsServiceConfigurer;
import org.springframework.security.oauth2.config.annotation.web.configuration.AuthorizationServerConfigurerAdapter;
import org.springframework.security.oauth2.config.annotation.web.configuration.EnableAuthorizationServer;

/**
 * 授权服务器配置
 */
@Configuration
@EnableAuthorizationServer
public class OAuth2AuthorizationServer extends
        AuthorizationServerConfigurerAdapter {

    @Override
    public void configure(ClientDetailsServiceConfigurer clients)
            throws Exception {
        clients.inMemory()
            .withClient("clientapp")
            .secret("123")
            .redirectUris("http://localhost:9001/callback")
```

```
            // 授权码模式
            .authorizedGrantTypes("authorization_code")
            .scopes("read_userinfo", "read_contacts");
    }
}
```

（5）定义一个资源服务器。

```
package com.cpf.authcodeserver.config;

import org.springframework.context.annotation.Configuration;
import org.springframework.security.config.annotation.web.builders.HttpSecurity;
import org.springframework.security.oauth2.config.annotation.web.configuration.EnableResourceServer;
import org.springframework.security.oauth2.config.annotation.web.configuration.ResourceServerConfigurerAdapter;

/**
 * 资源服务配置
 */
@Configuration
@EnableResourceServer
public class OAuth2ResourceServer extends ResourceServerConfigurerAdapter {

    @Override
    public void configure(HttpSecurity http) throws Exception {
        http.authorizeRequests().anyRequest().authenticated().and()
            .requestMatchers()
            .antMatchers("/api/**");
    }

}
```

完成上面的程序编写后，进行测试。

直接请求 localhost:8080/api/userinfo，发现请求资源没有权限，如图11-2所示。

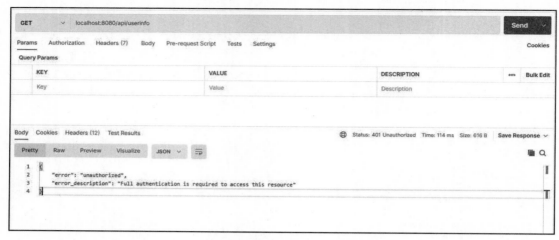

图 11-2　请求资源没有权限

获取授权码，使用浏览器请求 http://localhost:8080/oauth/authorize?client_id= clientapp& redirect_uri=http://localhost:9001/callback&response_type=code&scope=read_userinfo，如图 11-3 所示。

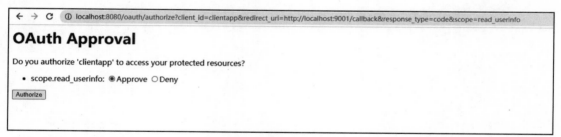

图 11-3　获取授权码

选中 Approve 单选按钮后，浏览器地址栏出现 http://localhost:9001/callback?code= 1cJbFb。这段地址前面的部分（http://localhost:9001/callback）无须关注，记下地址后面的部分 code=1cJbFb。

获取令牌，得到 access_token、token_type、expires_in 和 scope，如图 11-4 所示。

（6）访问用户数据。

获取用户数据如图 11-5 所示。

至此，我们已经完成了一次完整的 OAuth 2.0 访问资源的过程。

图 11-4　获取令牌

图 11-5　获取用户数据

11.4　JWT 使用

JWT，全称是 json web token，是目前比较流行的一种身份认证解决方式，下面开始本节内容的学习。

11.4.1　JWT 的介绍

令牌可以分为透明令牌（by reference token）和自包含令牌（by value token）两种。

- ☑ 透明令牌：随机生成的字符串标识符，无法简单猜测授权服务器如何颁发和存储。资源服务器必须通过后端渠道发送回授权服务器的令牌检查点，才能校验令牌的有效性。
- ☑ 自包含令牌：授权服务器颁发的令牌，包含关于用户或客户的原数据和声明，通过检查签名、签名的颁发者、签名的接收人或者 scope，资源服务器可以在本地校验令牌。通常的实现为 JWT。

JWT 定义了一种紧凑且自包含的标准，它将各个主体的信息包装为 JSON 对象。主体信息是通过数字签名进行加密和验证的。经常用 HMAC 算法或者 RSA 算法对 JWT 进行签名，所以安全性比较高。

JWT 由以下 3 个部分组成。
- ☑ 头部：用于描述关于该 JWT 的最基本的信息，如令牌的类型以及签名所用的算法等。将头部信息进行 Base64 编码生成 JWT 的第一部分。
- ☑ 载荷：就是存放有效信息的地方，在这一部分中存放用户信息、声明、权利、过期时间等。将载荷进行 Base64 编码生成 JWT 的第二部分。
- ☑ 签名：签证信息。典型的格式是 HMACSHA256(base64UrlEncode(header)+"."+base64UrlEncode(payload)+secret)。

header 和 payload 的信息是通过 Base64 进行编码的，可以进行反编码看到里面的信息。签名的计算需要指定一个密钥 secret，这个密钥只在服务端存储，所以只有服务端可以校验此 JWT 生成 Token 的有效性。

JWT 常用的格式为如下。

```
xxxx.yyyy.zzzz
```

11.4.2 JWT 的实践

我们来实践一下。

首先，在项目中添加依赖。

```xml
<!--jwt-->
<dependency>
    <groupId>io.jsonwebtoken</groupId>
    <artifactId>jjwt</artifactId>
    <version>0.7.0</version>
</dependency>
```

编写工具类，代码如下。

```java
package com.cpf.authcodeserver.util;

import io.jsonwebtoken.Claims;
import io.jsonwebtoken.Jwts;
import io.jsonwebtoken.SignatureAlgorithm;

import java.util.Date;
import java.util.concurrent.TimeUnit;

/**
 * @author 晁鹏飞
 * @date 2022/3/19
 */
public class JwtUtil {
    // 只在服务端存储的密钥
    private static final String secret = "asdfasdf";

    public static String createToken(String subject){

        String token = Jwts.builder().setSubject(subject)
        .setExpiration(new Date(System.currentTimeMillis() + 1000 * 60 * 60))
            .signWith(SignatureAlgorithm.HS256, secret).compact();

        return token;
    }

    public static String parseToken(String token){
        Claims body = Jwts.parser().setSigningKey(secret).parseClaimsJws(token).getBody();

        String subject = body.getSubject();
        return subject;
    }

    public static void main(String[] args) throws InterruptedException {
        String name = "教育";
```

```
        String token = createToken(name);
        System.out.println("生成的token:"+token);

        String srcStr = parseToken(token);
        System.out.println("解析值: "+srcStr);

    }
}
```

运行程序，发现程序会将"教育"这个内容作为载荷，生成 JWT 的 Token，运行结果如下。

```
生成的token:eyJhbGciOiJIUzI1NiJ9.eyJzdWIiOiLmlZnogrIiLCJleHAiOjE2MTk
2ODAxODh9.QET4RK2iVhpYFLu-qs13MbzDBEY5GjMBA9dHUAiBimY
解析值：教育
```

11.4.3　JWT 的使用场景

JWT 的使用场景如下。

- ☑ 授权：这是最常见的使用场景，能够解决单点登录问题。一旦用户登录，每个后续请求都将包含 JWT，允许用户访问该令牌允许的路由、服务和资源。因为 JWT 使用起来轻便、开销小，并且服务端不用记录用户状态信息（即无状态），所以使用比较广泛。正由于这种无状态的特性，还便于服务的水平扩展。
- ☑ 信息交换：JWT 是在各个服务之间安全传输信息的好方法。因为 JWT 可以签名，如使用公钥/私钥对，可以确定请求方是合法的。此外，由于使用密钥和有效负载计算签名，还可以验证内容是否未被篡改。

11.5　小　　结

本章介绍了微服务鉴权认证、安全方面的设计以及实践。首先从应用场景出发，分析了单体应用和微服务应用中的安全认证问题。然后详细介绍了 OAuth 2.0 的理论和实战，对常用的令牌 JWT 做了介绍，用一个小例子，讲解了自包含令牌 JWT 是如何使用的。让读者能在实际工作中达到安全使用服务的目的。